Urban Ethnic Encounters

A characteristic feature of cities has been their ethnic heterogeneity, caused by the arrival of job-seeking immigrants, refugees, temporary guest workers and tourists.

Urban Ethnic Encounters attempts to answer two leading questions: How does urban space structure the life of ethnic groups? How does ethnic diversity help to shape urban space? A multidisciplinary team of authors explore the various dimensions of the spatial organization of inter-ethnic relations in cities and countries around the globe. Unlike most ethnographies, in which authors write about the 'other' in faraway places, the majority of the contributors have studied their own society.

The case studies are from three different continents. Material is presented from diverse locations, such as the cities of Toronto, Philadelphia, Vienna, Beirut, Jakarta, Tehran, Osaka and Albuquerque, and the countries of Israel, Brazil and Taiwan, giving a unique opportunity for comparative analysis of ethnicity and spatial patterns.

Aygen Erdentug is an associate professor at the Department of Political Science, Bilkent University, Ankara, Turkey.

Freek Colombijn is a lecturer at the Department of Languages and Cultures of Southeast Asia and Oceania, Leiden University, the Netherlands.

Routledge Research in Population and Migration
Series Editors: Paul Boyle and Mike Parnwell

1 **Migration and Gender in the Developed World**
   Edited by Paul Boyle and Keith Halfacree

2 **Migration, Globalisation and Human Security**
   Edited by David T. Graham and Nana K. Poku

3 **The Geography of Trafficking and Human Smuggling**
   Khalid Koser and John Salt

4 **Population Mobility Among Indigenous Peoples in Australasia and North America**
   Edited by John Taylor and Martin Bell

5 **Urban Ethnic Encounters**
   The Spatial Consequences
   Edited by Aygen Erdentug and Freek Colombijn

Urban Ethnic Encounters
The spatial consequences

Edited by Aygen Erdentug
and Freek Colombijn

London and New York

First published 2002
by Routledge
11 New Fetter Lane, London EC4P 4EE

Simultaneously published in the USA and Canada
by Routledge
29 West 35th Street, New York, NY 10001

Routledge is an imprint of the Taylor & Francis Group

© 2002 selection and editorial material Aygen Erdentug and Freek
Colombijn; individual chapters the contributors

Typeset in Galliard by Taylor & Francis Books Ltd
Printed and bound in Great Britain by The Cromwell Press, Trowbridge,
Wiltshire

All rights reserved. No part of this book may be reprinted or reproduced
or utilised in any form or by any electronic, mechanical, or other means,
now known or hereafter invented, including photocopying and recording,
or in any information storage or retrieval system, without permission in
writing from the publishers.

British Library Cataloguing in Publication Data
A catalogue record for this book is available from the British Library

Library of Congress Cataloging in Publication Data
A catalog record for this book has been requested

ISBN 0-415-28085-0

Contents

List of illustrations	viii
Notes on contributors	x
Preface	xiv
Acknowledgements	xv

1 Introduction: Urban space and ethnicity   1
   FREEK COLOMBIJN AND AYGEN ERDENTUG

PART I
The macro-level analysis of urban ethnic encounters:
Enclaves and the zones of the cities   25

2 Residential segregation and neighbourhood socioeconomic
   inequality: Southeast Asians in Toronto   27
   JOE T. DARDEN

3 Places of worship in multicultural settings in Toronto   46
   CHANDRAKANT P. SHAH

4 The impact of local and regional planning on Arab
   towns in the 'Little Triangle', Israel   61
   THABET ABU RASS

5 Borders and boundaries in post-war Beirut   81
   DANIEL GENBERG

PART II
The meso-level analysis of urban ethnic encounters: The neighbourhoods of the cities 97

6 Perception and use of space by ethnic Chinese in Jakarta 99
HARALD LEISCH

7 Urban fear in Brazil: From the *favelas* to *The Truman Show* 109
CARMEN SÍLVIA DE MORAES RIAL AND MIRIAM PILLAR GROSSI

8 Ethnic consciousness arises on facing spatial threats to Philadelphia Chinatown 126
JIAN GUAN

9 Transcultural home identity across the Pacific: A case study of high-tech Taiwanese transnational communities in Hsinchu, Taiwan, and Silicon Valley, USA 142
SHENGLIN CHANG

10 To cross or not to cross the boundaries in a small multi-ethnic area of the city of Tehran 160
SOHEILA SHAHSHAHANI

PART III
The micro-level analysis of urban ethnic encounters: The streets and the squares of the cities 175

11 Repackaging difference: The Korean 'theming' of a shopping street in Osaka, Japan 177
JEFFRY T. HESTER

12 The appropriation of public space as a space for living: The Waterworld Festival in Vienna 192
HEIDI DUMREICHER

13 Contested urban space: Symbolizing power and identity in
   the city of Albuquerque, USA   209
   EVELINE DÜRR

14 Conclusion   226
   FREEK COLOMBIJN AND AYGEN ERDENTUG

   Index   246

Illustrations

Figures

2.1	Distribution of Southeast Asians and whites in Toronto CMA by socioeconomic quality of neighbourhoods, 1996	41
9.1	An American home environment in Hsinchu Science Park	146
9.2	Traffic congestion in Hsinchu Science Park	147

Maps

2.1	Distribution of Southeast Asians across Toronto CMA's census tracts, 1996	37
2.2	Distribution of white population across Toronto CMA's census tracts, 1996	38
3.1	The Buddhist community in Toronto	53
3.2	The Islamic community in Toronto	54
3.3	The Sikh community in Toronto	55
3.4	The Hindu community in Toronto	55
4.1	The Little Triangle region in Israel	62
4.2	The Little Triangle and its geopolitical significance	69
4.3	Arab and Jewish jurisdictions in the Little Triangle	71
6.1	Chinese space in Jakarta	103
8.1	Philadelphia Chinatown	127
10.1	Ethnic groups in Iran	161
10.2	Ekhtiarieh	163
13.1	The city of Albuquerque	212
13.2	The Old Town and the historic district of Albuquerque	213

Tables

2.1	Socioeconomic indicators for Toronto CMA by quality of neighbourhoods, 1996	40
2.2	Percentage and share of Southeast Asians and whites in Toronto CMA by socioeconomic quality of neighbourhoods, 1996	40

3.1	Hypothesized pathways by which dimensions of religious involvement influence health	47
3.2	Religion and health outcomes	49

Contributors

Thabet Abu Rass [Ph.D. University of Arizona, USA, 1997] is a lecturer in the Department of Geography and Environmental Development, Ben Gurion University of the Negev, Israel. His research and publications focus on local government studies, ethnic relations, and planning and regional development.

Shenglin Chang [Ph.D. University of California at Berkeley, USA, 2000] is currently an assistant professor in the Natural Resource Sciences and Landscape Architecture Department, University of Maryland, USA. She has developed and implemented innovative approaches to public involvement in environmental issues through civic arts, participation in community design and social–political activism. She is a co-editor, together with Randy Hester and Shih Wang, of *Living Landscape: Reading Cultural Landscape Experiences in Taiwan and America* (1999, in Chinese). A more recent electronic publication is 'A Study of the Environmental and Social Aspects of the Taiwanese and U.S. Companies in the Hsinchu Science-based Industrial Park' (2001) at www.nautilus.org.

Freek Colombijn [Ph.D. Leiden University, the Netherlands, 1994] is an anthropologist and historian specializing in Indonesia and lecturing at Leiden University, the Netherlands. He has co-edited, with Peter Boomgaard and David Henley, *Paper Landscapes: Explorations in the Environmental History of Indonesia* (1997) and, with J. Thomas Lindblad, *Roots of Violence in Indonesia* (2002). His interest in urban studies dates back to his dissertation, published under the title *Patches of Padang: The History of an Indonesian town in the Twentieth Century and the Use of Urban Space* (1994). Having completed a project on physical violence, he is currently carrying out fieldwork in Indonesia on the politics of road construction in the past and at present.

Joe T. Darden [Ph.D. University of Pittsburgh, USA, 1972] is Professor of Geography and Urban Affairs at Michigan State University, USA. His research interests are racial residential segregation and neighbourhood socioeconomic inequality in multiracial societies. He is the author of over a hundred publications including the co-authored book *Detroit: Race and Uneven Development* (1987). From 1997 to 1998 he was a Fulbright Scholar in the Department of Geography at the University of Toronto, Canada.

Heidi Dumreicher [Ph.D. University of Vienna, Austria, 1968] is a linguist, historian and urban researcher. She is the founder and an active member of the Oikodrom Institute for Urban Sustainability, Vienna, which is devoted to restoring the health and vitality of the modern urban habitat through research and active intervention. She is the editor of the quarterly *Oikodrom Stadtplaene* (since 1994). She has participated in various national and international meetings on sustainability in the city. She has also contributed as a journalist, on a variety of topics, to the department of science and society of the Austrian Broadcasting Company, as well as to leading popular European journals. Since 2002 she has been coordinator of the European Union research project SUCCES (Sustainable Uses Concepts for China Engaging Scientific Scenarios).

Eveline Dürr [Ph.D. University of Freiburg, Germany, 1990] is an assistant professor at the Institut für Völkerkunde, University of Freiburg, Germany. Her dissertation, on the rebellion of the Tzeltal (1712–13) in colonial Mesoamerica, is based on archive research and fieldwork in Guatemala and Mexico. Prior to her most recent fieldwork in the city of Albuquerque in the southwest of the USA, she produced a monograph on cultural change in a Zapotec community in Mitla in the state of Oaxaca, Mexico.

Aygen Erdentug [Ph.D. Hacettepe University, Ankara, Turkey, 1980] is an associate professor at the Department of Political Science and Public Administration, Bilkent University, Ankara, Turkey. She is a sociologist-turned-social anthropologist whose country of specialization is Turkey. Her fields of interest and publications cover urban and rural sociocultural change, the family, kinship and marriage, the anthropology of work and organizations, political symbolism and ethnicity. She is currently involved in documenting the development of Turkish social anthropology, as well as exploring the various dimensions of cross-cultural marriages in Turkey.

Daniel Genberg is a Ph.D. candidate at the Department of Social Anthropology, Stockholm University, Sweden. His research and forthcoming publications (including chapters in *Contesting Good Governance* for Curzon Press and in *Crisis and Memory: Dimensions of Their Relationship in Islam and Adjacent Cultures* for Oxford University Press) focus on the phase of reconstruction in Beirut and more generally on urban space and planning.

Miriam Pillar Grossi [Ph.D. University of Paris V (René Descartes), France, 1988] is a professor ('professora adjunto') of anthropology and Director of the Centre of Philosophy and Humanities at the University of Santa Catarina (UFSC), Brazil. She is also the editor of the *Revista Estudos Feministas*, published in Brazil, and a co-author, with Analba Brazao, of *Histories to Tell: a Portrait of Physical and Sexual Violence Against Women in* the City of Natal, Brazil (2000, in Portuguese). Her research and publications cover gender studies, violence, history of anthropology and urban anthropology.

Jian Guan [Ph.D. Oklahoma State University, USA, 1996] is an assistant professor in the Department of Sociology at Pennsylvania State University, USA. Her research focuses on urban community, ethnic studies, cross-cultural studies and public health. Her publications in the USA have been on the acculturation and marginalization of Chinese ethnic groups, along with the impact of cross-cultural interactions on value changes among Chinese students. She has also published (in China) on value changes among the Chinese minorities in the process of modernization and the economic development in Hainan Island and (in India) on the religious aspects of the Dais in Xipsuang Banna.

Jeffry T. Hester [Ph.D. University of California at Berkeley, USA, 1999] is Associate Professor of Sociocultural Anthropology in the Asian Studies Program, Kansai Gaidai University, Osaka, Japan. His dissertation is on place-making practices and national identity among Koreans and Japanese in an Osaka neighbourhood. In this context, he has written a chapter in the publication *Koreans in Japan* (2000) concerning ethnic education for Korean children in Osaka. His research interests include ethnicity and national identity, citizenship, cultural performance, gender and youth.

Harald Leisch [Ph.D. University of Trier, Germany, 1994] is a scientific coordinator for a collaborative research programme of the University of Hohenheim (Germany) in Vietnam. His dissertation is in the field of medical geography, on the medical infrastructure and population development in Northern Thailand (published in 1994). His current research project focuses on the social aspects of urban development, urban ethnicity, gated communities, and regional development in Indonesia. To this end, he received a grant (1999–2001) from the German Research Association (DFG) for fieldwork on middle-class life in the new towns in Jabotabek, Indonesia.

Carmen Sílvia de Moraes Rial [Ph.D. University of Paris V (Sorbonne), France, 1992] is a professor ('professora adjunto') in the Department of Anthropology at the University of Santa Catarina, Brazil. She is Director of the Post-Graduate Program in Anthropology, as well as being the editor of *Ilha – Revista de Antropologia*. Her research and publications (mostly in Portuguese) focus on culture and globalization, visual anthropology and anthropology of food, and include two forthcoming books: *Anthropology of Food in Brazil* and *Gender Studies* (both in Portuguese).

Chandrakant P. Shah [MD, FRCPC, MRCP(Glas), SM(Hyg)] is a professor and Director of Development and Communication in the Department of Public Health Sciences. He is also cross-appointed in the various departments of the Faculties of Medicine, Social Work and Nursing at the University of Toronto. His publications and research projects cover aboriginal health and the health of marginalized groups, various health issues related to social class, to unemployment, to equality in employment, to spirituality, to hunger and to homelessness. He has received numerous honorary memberships or titles for his contributions to and his advocacy for these disadvantaged populations. His book *Public Health and Preventive Medicine in Canada*, also translated into French, is widely used as a textbook in medical schools and in other health science disciplines across Canada.

Soheila Shahshahani [Ph.D. Graduate Faculty of the New School for Social Research, New York, USA, 1981] is an assistant professor at Shahid Beheshti University in Tehran, Iran. Her research in urban anthropology covers the cities of Kashan, Varameen, Esfahan and Tehran. She has also carried out fieldwork on the Mamassani, a pastoral nomadic population in Iran, and in the grotto village of Meymand in central Iran. Her publications include three books: *The Four Seasons of the Sun*, *A Bibliography of Iranian Women in Social Sciences, Humanities and Arts* and *A Pictorial History of Iranian Headdresses* (all in Persian). She is editor of the new journal *Anthropology* (2001, also in Persian).

Preface

Twelve of the chapters in this volume (Chapters 2 to 13) are revised versions of a selection of papers presented at the panel 'The Spatial Consequences of Urban Ethnic Diversity' that convened at an Inter-Congress of the International Union of Anthropological and Ethnological Sciences (IUAES), held at Beijing on 24–28 August 2000.

Taken together, the chapters search the various dimensions of urban interethnic relations as reflected in the spatial organization of cities around the globe. Unlike most ethnographies, in which authors write about the 'other' in faraway places, the majority of the authors in this volume have studied their own society. This 'anthropology at home' gives the analyses a refreshingly intimate and direct touch. However, the term 'anthropology' can be misleading, since the authors actually come from diverse backgrounds, such as the fields of geography, landscape architecture, social demography, medicine, sociology and, indeed, social or cultural anthropology. This multidisciplinary background enables the adoption of a rich variety of methodological approaches, thus pointing out the many directions for possible research in the future. Despite their national and disciplinary variety, the authors of this book have adopted an 'actor-oriented' approach and share the assumption that 'urban space' is not a given, but a social construct.

It was not possible to arrange the chapters geographically due to the restrictions of the methodology of social sciences concerning the analysis of any given data. Hence the 'Introduction' has been based on a classification clustered according to the level – that is, the macro-, the meso- and the micro-levels – being analysed by the author of each chapter. This proved to be a difficult task because of the blurred demarcations between the different levels of analyses and the multiple, if not innovative, approaches of the authors.

We would like to thank the Chinese Association of Urban Anthropology for organizing the Inter-Congress of the IUAES and for their gracious hospitality, enabling the authors of this volume to meet and discuss their papers in detail. We also wish to extend our appreciation to the authors who have patiently put up with our requests during the process of editing and preparing the chapters for this book.

Aygen Erdentuğ and Freek Colombijn

Acknowledgements

The authors and publishers would like to thank the following for granting permission to reproduce material in this work:

Brill Academic Publishers, The Netherlands, for the figure 'Distribution of White People Across Toronto CMA's Census Tracts, 1996', from the *Journal of Developing Societies*, Vol. XVI(2), 2000, pp. 245–270 (map 2.2 in *Urban Ethnic Encounters*).

Dietrich Reimer Verlag GmbH, Berlin, for the figure 'Map of Albuquerque', by Eveline Durr, from *Feldforschungen*, 2nd edition, edited by Hans Fischer, 2002 (map 13.1 in *Urban Ethnic Encounters*).

Elsevier Science, for the table 'How Religion Influences Morbidity and Health', by J. S. Levine, from the journal *Social Science and Medicine*, Vol. 43 (5), p. 858, 1996 (table 3.1 in *Urban Ethnic Encounters*).

Every effort has been made to contact copyright holders for their permission to reprint material in this book. The publishers would be grateful to hear from any copyright holder who is not here acknowledged and will undertake to rectify any errors or omissions in future editions of this book.

1 Introduction

Urban space and ethnicity

Freek Colombijn and Aygen Erdentug

If we accept the arbitrary, yet widely accepted, claim that the Chicago School is the progenitor of urban sociology (Hannerz 1980: 20), then 'ethnic diversity' can be seen to have been an important component of urban studies since its conception. About eighty years ago, Robert Ezra Park and his colleagues at the University of Chicago published their pioneering studies on the city. In their view, the city was made up of different groups of people, defined in terms of social class and ethnic background, with each group finding a niche in the city in which to work, to live or to spend their leisure time. Ethnic groups tended to live in certain quarters and economic functions were clustered in certain areas. Expanding ethnic groups or booming economic functions invaded the territory of other groups or functions, as well as succeeding each other in one particular spot. Different groups competed for space – for instance, street gangs (mostly originating from specific ethnic groups) were organized territorially – but also established symbiotic relationships to share certain spots in town (Park, Burgess and McKenzie 1967 [1925]; cf. Eriksen 1993: 18–20; Hannerz 1980: 19–58). Louis Wirth, another Chicagoan, asserted that the city has historically been a 'melting-pot', which 'has brought together people from the ends of the earth *because* they are different and thus useful to one another' (Wirth 1938: 10). Hence, inter-ethnic relations are more likely to develop in cities than in villages, because the diversity of services and opportunities offered by cities attracts a larger variety of people than a village economy. Moreover, statistics suggest the probability that the bigger the population, the higher the number of different ethnic groups. Wirth was pessimistic about the superficial, anonymous and transitory nature of urban relationships. However, later research – by Herbert Gans (1982 [1962]) on the Italians in Boston's West End – showed that friendly, closely knit communities developed in immigrant neighbourhoods, partly because communal life mostly took place on the street. These neighbourhoods showed considerable residential stability.

Chicago remains as one of the cities where ethnic ('racial') segregation has persisted, if not increased, in the course of the twentieth century (McCarthy 1999). Meanwhile, a number of cities like Jerusalem, Beirut, Belfast, as well as Cape Town, Johannesburg and other South African cities – perhaps also Berlin, before the unification of Germany – have become deeply divided with ethnic borders, or are struggling to overcome the former internal ethnic demarcations of urban living. Other cities in North America and throughout the world lack total ethnic segregation, but are identified by spatial–ethnic mosaics. Envy, mistrust, avoidance and violence characterize the ongoing inter-ethnic relationships. However, extending Jane Jacobs' argument (1964 [1961]) that diversity gives life to cities, from the built-up environment to the social structure, we believe that inter-ethnic encounters in cities are, on the whole, potentially enriching experiences. Since Park's time, the relationship between space and ethnicity has become all the more important, due to a number of developments in both global space and the global migration of ethnic groups.

All cities have now become part of the global network society, which is dominated by flows of information, images, people, capital, technology and goods. Nodes in the network develop where a large number of flows converge (in what Manuel Castells calls the 'space of flows'). Places that are not suitable or not connected to the network are bypassed, being left to live their own unobtrusive lives in the 'space of places'. The division between 'space of flows' and 'space of places' cuts across the former demarcation of the First World and the Third World. For example, stock exchanges, internet cafes and arms markets in less developed countries are nodes in the network, while a ghetto in New York can be disconnected from it (Appadurai 1996a; Castells 1996; Nas and Houweling 1998).

One result of the rise of the network society has been the polarization seen among the social classes of the cities. The upper class, consisting of businessmen, managers, professionals and information technology people, has swollen up, while the section of the middle class working in the manufacturing sector has decreased and the lower class that supports the upper class by cleaning, catering and personal services has increased. The non-skilled and low-paid lower-class jobs have been consigned, predominantly, to the new immigrant labour force. This polarization in the social structure, however, is still being contested and it is, at least, not encountered in all global cities. One spatial result has been the gentrification of dilapidated old neighbourhoods, opening them up for use by the new upper classes (Hamnett 2001; Phillips 1996: 422–3; Sassen 1991). A more recent morphological trend in urbanization is cities growing and blending together into mega-urban regions (e.g. Lo and Yeung 1996; McGee and Robinson 1995).

Since the first half of the twentieth century, new flows of migration or 'ethnoscapes' (Appadurai 1996a: 33) have emerged, complementary to the American Dream that drew migrants to North America. The Third World has gone through a process of decolonization that is changing the ethnic balance of power in these countries. Most communist states have collapsed and, with those regimes becoming defunct, the policy of keeping firm control over the movement of people from various ethnic backgrounds has also been abandoned. In the meantime, the nation-states in the West have experienced massive immigration from all over the world. In the case of Europe, this was initially due to the subjects of former colonies seeking a better life in the homeland of their previous rulers. More recently, the influx to the West has been in the form of 'guest workers' from economically disadvantaged countries and refugees seeking asylum from political unrest and oppression, if not genocide or ethnocide. Citizens of developing countries such as the Philippines, Pakistan and Bangladesh have consented to working as contract labourers in countries like Japan, Hong Kong and Saudi Arabia which are experiencing a labour shortage. Most of these immigrants have settled in the more cosmopolitan cities of the world. With globalization and the transnational movement of people, foreign tourists have also had a growing impact on city life (Appadurai 1996a; Castels 2000; Eriksen 1993: 1–3).

The main changes with regard to migration, over the course of the twentieth century, have been in the nature of the relationships between the migrants and their respective regions or countries of origin, replacing the situation where immigrants were cut off from their home town or country with each shift in the waves of migration. Time–space compression by technological innovations in transportation and communication has made it much easier for migrants to maintain contact with their respective places of origin. The repeated exposure to and remembrance of their homes, by air travel and through the mass media, has made migrants more aware of their ethnic identity. They can remain emotionally and politically more involved with their place of origin than with their new place of residence, even to the point that a Punjabi living in Toronto and holding a Canadian passport sends money and weapons for armed resistance in Khalistan. This 'long-distance nationalism' must have, to put it mildly, an impact on that person's integration process in Toronto. Nation-states can no longer assimilate within their national boundaries communities experiencing a diaspora as easily as they used to do at the beginning of the twentieth century (Anderson 1992; Hannerz 1996; Harvey 1990: 240–59, 302–7; Tajbakhsh 2001: 5, 8).

The variations in urban development and the new flows of migration with the concomitant new ethnic composition of cities have created ample reasons to revive the old subject of the relationship between urban space and ethnicity. People from various ethnic backgrounds do not spread out in cities at random; they create a patterned spatial distribution. It is this topical issue – namely, the spatial consequences of urban ethnic diversity – that this book aspires to explore. It deals with the mutually interactive relationship between ethnicity and urban space in several countries. In other words, the relationship between ethnicity and urban space is a two-way process. To begin with, space has an impact on how people from different ethnic groups (besides being of a certain class, age and gender) lead their lives. Conversely, since the members of the ethnic groups have, by definition, a different cultural background and develop dissimilar economic activities, they tend to utilize the urban environment in diverse ways, having a different impact on the transformation of urban space. As Winston Churchill said when he re-opened the House of Commons after World War II: 'We shape our buildings and then they shape us' (quoted in Phillips 1996: 462). The well-known statesman, as usual, probably oversimplified to popularize an issue, giving the erroneous impression that the impact of the environment on people is deterministic. In reality, buildings (and, for that matter, urban space) cannot be solely accepted as agents producing a result. There is a multiplicity of interactive factors that determine what we build or set aside as space. In return, while these maintain certain livelihoods and beliefs, they can also either force – again under the influence of certain factors – people to be creative or frustrate them, thus altering their ways of life.

Therefore, two main questions determine the orientation and the scope of this volume: How do members of ethnic groups adjust their lives to the existing urban space and, conversely, how does ethnic diversity contribute to creating the shape of urban space? The authors elaborate these central questions through a range of more particular questions, such as: Why do some ethnic groups feel the need to be segregated, while others manage to blend into the 'melting pot'? Is spatial allocation spontaneous or enforced? Where and how do segregated ethnic groups have the opportunity for interaction? Are there any sections of the city that they claim exclusively for themselves? Which quarters are forbidden, if not dangerous, zones for other ethnic groups?

This book attempts to fill a gap in this field of interest. Studies of ethnicity in general (e.g. Barth 1969; Dekker, Helsloot and Wijers 2000; Eriksen 1993; Jenkins 1996, 1997) or urban ethnicity in particular (Cohen 1974; Eriksen 1993: 131–9; Guldin 1985; Hannerz 1980; Rogers and Vertovec 1995) disregard spatial aspects completely or almost completely. The ethnic encounters seem to take place in a void. Studies of urban morphology and urban social structure have done better in this respect and have paid much attention to ethnicity. However, even these urban studies tend to take ethnic groups as fixed and permanent categories, thus ignoring a whole body of anthropological and theoretical literature on the subject (see, however, Tajbakhsh 2001). Moreover, most urban studies are heavily 'skewed' to present the English-speaking world, especially the United States and the United Kingdom (e.g. Gottdiener 1985; Paddison 2001; Phillips 1996). Such studies run the risk of making unjustified generalizations based, most often, on specific American circumstances. The authors 'forget' to inform the readers that their conclusions apply to the United States only, probably because they tend to take it for granted that readers are only interested in the United States. Worse than that, they miss the point that these allegedly idiosyncratic urban American problems are widely encountered in the world. One example of American bias is the exaggerated concern about 'race', which is central to the discourse in US society but, more often than not, not so much of an issue elsewhere. British scholars do not fare any better in this respect, preferring to look at urban realities through the 'windows' of colonialism and post-colonialism.

This volume attempts to break with this tendency through global comparisons, presenting cases from three different continents. These global comparisons enhance the applicability of the findings, which can contribute to a better understanding of the integration of ethnic groups in urban settings. In the concluding chapter, we use the main findings to make a comparison of the cases presented. Given the limited number of studies, these findings should be seen as initial conclusions or hypotheses to be elaborated and tested through further research. Since the two key concepts in this book are 'urban space' and 'ethnicity', we will discuss them first, before presenting the cases from different countries.

Urban space

The starting point of our theoretical excursion into the debate on 'space' is the human ecology approach of the Chicago School that has continued to inspire scholars for decades. In the seminal article from 1925, Ernest Burgess discussed, as a whole, the ethnic–spatial relationships at the macro-level of a city. Burgess argued that the ideal type of city developed, roughly, in concentric circles. In other words, the central business district was at the heart of the city, surrounded by a so-called 'zone in transition' that was invaded from the centre outwards. Then there was the second outward zone of immigrants, along with a residential area of single-family dwellings. Residents of parts of the transitional zone that had deteriorated would move out to the various residential shells further away from the centre. The competition over space was mediated through the price for land or, as suggested by Goldberg and Chinloy (1984: 103–10), through higher rents. Since people from various ethnic groups could afford to pay different prices – that is, they had different buying power – to occupy the most favourable spaces, ethnic groups tended to cluster in different circles of the city (Burgess 1967 [1925]). Variations on the model of development in concentric circles were put forward by Hoyt, Harris and Ullman (Phillips 1996: 423–7) and, for Southeast Asia, by Terry McGee (1967).

Other work from the Chicago School supported the macro-analysis of Burgess at the neighbourhood level (Park 1967 [1925]: 7–12). Louis Wirth, for instance, described in *The Ghetto* (1928), after a long introduction on the emergence of the ghetto in Europe, how a Jewish neighbourhood gradually developed in Chicago as more and more such immigrants settled in one particular area of the city. Community life was organized through synagogues, burial societies, religious schools and other institutions. The price to be paid for neglecting the relationships between Jews and outsiders and focusing instead on the community life within this ethnic quarter was impeding the integration process into the greater society. At the initial stage of the migration history of an individual, it was instrumental to seek protection in the ghetto – that is, within a particular ethnic group. Yet at the next stage – during integration and later assimilation into the cultural mainstream, along with the attainment of some prosperity – the ghetto lost its purpose. Successful residents left the ghetto and moved to more affluent neighbourhoods. The ghetto remained intact, however, because new immigrants succeeded those who moved out.

Through micro-level analysis of a bounded space (that is, the street, square or market), the ethnographer can study face-to-face social interaction between people of various ethnic backgrounds. A fine, classic example from Chicago at this level is Paul G. Cressey's *The Taxi-Dance Hall* (1932). The taxi-dance hall was a kind of dancing school, where customers paid girls, known as "taxi-dancers", to dance with them. Cressey focused on the rules of conduct and the economic dynamics at work in this setting. In the most popular halls, the most attractive girls were willing to dance with men from those ethnic groups that they considered appealing. Men from the less appealing ethnic groups – in the perception of the girls, Filipinos in particular – had to make do with the less attractive girls, or frequent the less popular dance halls. Not unlike Burgess' city of concentric circles based on the price of land, the spatial pattern of where men of various ethnic groups could be found was mediated through the market, in this case through the price of the girls and the amount that the men could afford to pay. An important theme throughout Cressey's study is how ethnic groups are categorized and how they are ascribed certain values by outsiders. The latter theme is important because Cressey transcends the naive, essentialist image of an ethnic identity being a fixed identity. Insights of the Chicago School, such as the idea of competition for land, mediated by the land price, between ethnic groups as a mechanism in organizing space and the migration history of neighbourhoods are still very useful today. When Burgess' model is tested outside America, one point of criticism that comes up is that he has incorrectly assumed that land is always a commodity (Hannerz 1980: 28). Other weaknesses of the Chicago School were its atomistic view of society and its one-dimensional view of space as simply the physical and built environment.

A breakthrough, overcoming the weak points of the Chicago School, came from Henri Lefebvre in his *La production de l'espace*, originally published in 1974. His work is referred to by many, but read by few. This is because of Lefebvre's highly abstract thinking, his tiringly mystifying style of writing (with flowery, but imprecise expressions, rhetorical questions and exclamations like 'L'espace!' – with which he begins his book) and the lack of concrete examples and references to other authors. Even as ardent an admirer as Gottdiener admits that he 'is never quite sure how to replicate his [Lefebvre's] reasoning' (Gottdiener 1985: 157). Lefebvre's main postulate is that 'space' is not an empty, meaningless, geometrical container to be filled in by human activities, but a social product in itself. In other words, space is produced by the social relations of production and reproduction, while space itself produces social relations. It is simultaneously a product, a means of production, a part of the social forces of production and an object for consumption. Space has a strong symbolic value and is 'inscribed' with meaning. Lefebvre distinguishes between *espace perçu*, *espace conçu* and *espace vécu*: The 'perceived space' (*espace perçu*) is the spatial reality (*la pratique spatiale*) of a materially constructed environment. The 'conceived space' (*espace conçu*) consists of what is represented (*les représentations de l'espace*) by planners, administrators, scientists and technocrats. The 'experienced space' (*espace vécu*) is formed when the space is supplied with images and symbols (*les espaces de représentations*) by the ordinary users of urban space (Lefebvre 1986 [1974]). Lebevre's claim that 'space is a social product' has been adopted by scholars from various backgrounds (e.g.

Boyer 1983; Castells 1975; Colombijn 1994; Gottdiener 1985; Gottdiener and Feagin 1988; Kilian 1998; Öncü and Weyland 1997; Philips 1996: 416). The positive elements in Lefebvre's work are his unfailing focus on space as an object of study and his attention on the non-physical and symbolic aspect of space. The morally weighted distinction between *espace conçu* and *espace vécu* is, in our view, less useful. Both concepts are concerned with the meaning people attach to physical space. Whether this meaning comes from professionals ('baneful' according to Lefebvre) or from ordinary people ('artful') does not matter much.

The atomistic view of society that formed the backbone of the Chicago School became the main focus of attack for the 'new urban sociology' (Gottdiener and Feagin 1988; Phillips 1996: 432–4), which stressed the role of power in the allocation of urban space. Marxist scholars (including Lefebvre) have mostly been associated with this new urban paradigm. They noticed that the human ecology approach failed to provide explanations for urban riots and the explosion of racial violence. The capitalist mode of production and the inevitable conflicts between social classes seemed to be at the root of such urban tensions. A theoretical problem for the Marxist scholars was generated from the fact that the urban grassroots movements were not concerned with the means of production, but fought over issues related to consumption. These consisted of problems such as affordable housing, participation in neighbourhood decision-making, the fate of housing blocks facing demolition and the distribution of municipal facilities. Hence, class did not create a rallying point for social action in the city. The Marxian urbanists tried to close the gap between the presumed class source of existing social contradictions and the empirically observable non-class basis of urban movements by insisting on a separation of the spaces for urban workplaces and for community residences (Tajbakhsh 2001: 2, 13–20).

In our view, the attempt to explain urban conflict in a Marxist framework is a bit forced: empirical facts are subjugated by theoretical rigidity, when it should be the other way round – that is, empirical tests should be the master in theory formation. A strict Marxist view is, in particular, no longer tenable in the present post-industrial economy of the Western countries. It is also not fully adequate for many non-Western societies, where many of the premises of a capitalist economy would not find support. For instance, there are situations in which land is not a commodity, because tradition prescribes that landownership is to be regulated by kinship rules and that land should not be for sale (Colombijn 1992).

A corollary of the Marxist view was the reductionist view of the state as being a continuation of the capitalist class (Tajbakhsh 2001: 19). This view also seems to be rather forced and depends on institutional arrangements. For example, landownership forms a very powerful resource in the United States, where the landowners hold a key position in negotiations about the use of urban space. In states with a weak judiciary, such as Indonesia, the value of land is eroded by the insecurity associated with legal titles. In such places decisions taken by landowners can be overruled by a coalition of the state and military which, to a considerable extent, is at liberty to act arbitrarily. We therefore prefer to consider the state as a *strategische Gruppe* (strategic group) with its own financial interests and political agenda, and its own peculiar means of planning, surveillance and bureaucratic rationalization. The state is not a unified entity, but is itself divided into competing interest groups (Evers and Schiel 1988; Tajbakhsh 2001: 20).

Another frontal attack on the atomistic view of the Chicago School came from the humanistic planners. Herbert Gans (1982 [1962]) argued that urban neighbourhoods could develop into tightly bound communities. But this had already been maintained by Louis Wirth in his study of the ghetto, a view that preceded his later notion of the city-dweller as leading an isolated way of life (Wirth 1928, 1938). Burgess' descriptive model of the city made up of concentric zones was also elaborated and transformed into a prescriptive model, with zoning as a planning technique to reduce urban chaos. This idea of zoning was detested and contested by Jane Jacobs (1964 [1961]) in her well-known plea for diversity (that is, mixed functions, short blocks, the mingling of old and new buildings and a dense concentration of people) as an indispensable condition for liveable cities. Urban planning, according to Jacobs, must prevent the winners in the competition for urban space casting out the losers (see also Appleyard 1981). The plea for diversity can be rephrased as the notion that ethnic diversity can, *in principle*, increase the quality of urban life, provided that social conflicts between ethnic groups can be resolved.

The post-modernist critique of Marxist urban studies is concerned with the semiotics of space. Space is imbedded in cultural values. Space is no longer seen as bounded, but is considered to be ambiguous with undecidable, indeterminable boundaries. Each person has her or his own perception of urban space and these spaces, in plural, overlap. Space is also gendered. Cultural representations (such as museums, imaginary architecture, Disney-like public spaces) are important elements in cities that may attract foreign investors. Kian Tajbakhsh suggests using the word 'spacing' instead of 'space', in order to express better the fluidity of space. Spacing, in other words, 'points to a space of becoming, rather than being' (Tajbakhsh 2001: 20–1, 28). The post-modernist view supersedes the older tradition of cognitive maps (Gould and White 1992; Lynch 1960; Phillips 1996: 477), with post-modernist criticism not only enriching but also elaborating Lefebvre's notion of *espace vécu*. However, we prefer to think in terms of one physical space (and not 'spaces' or 'spacing') and multiple representations.

Recapitulating our discussion of space so far, it is clear that we have preferred an eclectic approach in the theoretical basis of the volume. The starting point for producing this book had been an empirical curiosity (what are the spatial consequences of urban ethnic diversity today?) and not a theoretical position. Thus, each chapter leans more towards one or the other of the possible theoretical orientations, depending on the specific issue concerned and the disciplinary background of the author. What is common throughout the book, however, is the treatment of space as 'a social product' of the interaction between numerous actors who are, to a considerable extent, formally organized as groups and institutions. These actors – such as administrators, landowners, urban planners, owners or tenants of shops and office space, residents, grassroot action groups and squatters – have different relationships with any given space. There are multiple, fluid and overlapping representations of space, but there is only one physical space. The balance of power between the actors determines which party, or coalition of parties, potentially has the most impact on the production of space. The power of each actor depends on the resources at their disposal, while the strength of a certain resource depends on the institutional arrangements current locally. Given the fact that ethnic groups have often, although not necessarily, a different power base, their impact on the urban space tends to be unequal.

Ethnicity

Existing studies on the concept of 'ethnicity' suggest a triple mode of analysis – that is, the adoption of either a 'primordialist', an 'instrumentalist' or a 'constructivist' approach to examining its dimensions (Banks 1996; de Vos 1996: 15–20).

In the 'primordialist' approach (Geertz 1973), ethnic identity markers such as descent, language or religion are treated as 'given', deeply rooted and almost insusceptible to change; ethnicity mainly satisfies the individual's psychological need for self-identification. In urban studies, this idea of an essential core was reflected in Robert Park's image of the city as a 'mosaic of little worlds which touch but do not interpenetrate' (quoted in Tajbakhsh 2001: 9). In Geertz's view, primordialist sentiments formed an obstacle to state formation and modernity. Arjun Appadurai has pointed out that ethnicity, in the primoridialist view, connotes 'backwardness'. More importantly, he argues convincingly that primordial attachments do not precede formation of a state, but emerge as a result of state formation and the local interpretation of national and global developments (Appadurai 1996b; see also Verdery 1994: 44–6).

In reaction to the rather static primordialist approach, the 'instrumentalist' approach focuses on the questions of why and under what conditions ethnicity becomes a means to an end. It pays to play the trump card of ethnicity when access to certain rights or quotas provided by affirmative action policies – such as the allocation of jobs and housing, reserved seats in parliament or tuition waivers at school – are linked to ethnic affiliation. Ethnic identity formation is the result of the dynamics of elite competition. While some elite groups find it instrumental to formulate an ethnic identity, other elite groups have an interest in downplaying ethnic distinctions. Therefore ethnic identity formation is reversible. Some latent ethnic groups never make the transition to a subjectively perceived ethnic community (Brass 1991: 13–40). In the worst case, however, the construction of ethnicity may result in 'ethnic cleansing' and the total segregation of ethnic groups that used to be intermingled (Oberschall 2000).

An example of instrumentalist studies *avant la lettre* is the series of studies undertaken in the Copperbelt towns of Zambia in the 1940s, 1950s and 1960s.

These studies suggest that the people became more aware of their ethnic roots because of the high frequency of inter-ethnic encounters in towns and cities. The outcome of this realization had been the endeavours of the more recent migrants to the cities to 'retribalize' and to 'overcommunicate' their ethnic identity (Eriksen 1993: 20–2; Hannerz 1980: 119–62). It follows that not only are ethnic relationships more complicated in cities, but that the spatial consequences of inter-ethnic relations also tend to be more important in such a habitat since land in cities is scarce and the stakes in urban space are higher than in rural space.

The instrumentalist approach may explain why ethnicity comes into play, but cannot explain how ethnic boundaries are formed. The 'constructivist' approach, instigated by Fredrik Barth (1969) in his seminal collection entitled *Ethnic Groups and Boundaries*, fills this void. Ethnic boundaries are the outcome of social interaction between groups. Groups identify themselves *vis-à-vis* other groups and need the contrast, or mirror, of other groups to point out their own particular qualities and those of others. The criteria that are used to categorize 'Us' and 'Them' are not objective differences, but features that the actors themselves regard as significant. Ethnic boundaries in this view are not given, but constructed, manipulated, subject to change and situational. Various groups can interpret boundaries differently.

Barth did away with the content of ethnicity – the criteria that marked the boundaries – as 'cultural stuff' (Barth 1969: 15). In a revision of his own work two decades later, Barth admitted that the symbolic signs used to set the boundaries are chosen less haphazardly than he had assumed. But he reiterates his point that the focus must be on the boundaries and not on cultural content (Barth 1994: 16). But then he 'threw out the baby with the bath water', for if only the boundaries and not the content were to be taken into account, it would be difficult to explain why ethnicity can invoke such potent and stubborn emotions. This brings us back to the primordial attachments. Part of the 'cultural stuff' is actually a new invention of tradition. The three approaches are complementary, rather than contradictory, and ideally should be integrated (Appadurai 1996b; Cohen 1974; Eriksen 1993: 3–12, 18–35, 92–6; Jenkins 1996: 90–103; Jenkins 1997; Schefold 1998).

In his 1994 re-evaluation of his earlier work, Fredrik Barth suggests distinguishing between three different levels of analysis in dealing with ethnicity. The macro-level deals with how state policies influence the making of ethnic boundaries; the meso-level deals with leadership and rhetoric that creates collectivities; and the micro-level focuses on the role of interpersonal interaction in ethnicity. This demarcation is meant to be a heuristic tool (Barth 1994: 20–31), assisting one in further discoveries.

Like Barth, we believe that this three-level analysis is an effective tool, but it must be modified somewhat to meet the requirements of studying urban space. At the macro-level of analysis, the spatial consequences of ethnic diversity have to be examined for the whole of the city or region. Residential segregation and ethnic enclaves can only become visible at this level. The meso-level should focus on the life of ethnic neighbourhoods, while the micro-level undertakes the analysis of bounded space – that is, the street, the square or the market. The study of each level assumes the existence of the other levels as well. So, when exploring at the micro-level, one should be aware of the overall pattern of the city. Conversely, when taking a bird's-eye view of the macro-level one should keep it in mind that inter-ethnic encounters look different on the ground.

The case of Kota Ambon: revealing the fluidity of spatial and ethnic boundaries

Ambon City (Kota Ambon) lies on the island of Ambon and is the capital of the Indonesian province of the Moluccas (Maluku), an archipelago. In the sixteenth and seventeenth centuries, the Moluccas were famous for the trade in spices. Traders from the sultanates of Ternate and Tidore, Portugal and the Netherlands introduced Islam, Catholicism and Protestantism to Ambon. After the spice trade declined, the island had little to offer in economic terms, so government service became the most important and reliable source of income. Recruitment to the army, civil service, business, universities, professions and criminal networks was based on religious affiliation. Villages were either Muslim or Christian, *en masse*, with inter-village peace being maintained by a ritual alliance called *pela*, which had developed between specific villages of different religions (Mearns 1999; Sidel 2001).

Spatial divisions in Ambon City overlapped and its residents had multiple identities. However, the connection between one's place of residence and one's identity was nonetheless weak. Since the wards were, as a rule, either Muslim or Christian, the overall spatial pattern of Ambon City reflected a religious mosaic. During most of the 1990s, the only noteworthy difference that existed was the one between the original Ambonese – whether Christian or Muslim – and recent immigrants. The newcomers had come from various regions of Indonesia, but were locally known by a generic name – the 'Butonese' – as if they had all come from the island of Buton. The 'Butonese' were Muslims and they slowly tipped the religious balance on the island to the Islamic side.

In the view of the original population, as was also the case with many of the immigrants, Ambon City was a place full of supernatural forces that deserved respect. (The belief in spirits is not seen as conflicting with the beliefs of the monotheist world religions.) For instance, when the government encountered difficulties during the development of the roads – neglected local spirits could seriously hamper road construction – this was immediately taken as a failure to pay due respect to the local spirits. Only the descendants of the people that used to dwell in the rural princely domains that preceded the foundation of Ambon City had the ritual knowledge necessary to deal with the local spirits. The original population can therefore be divided into two: the descendants of the former inhabitants and the later settlers from surrounding villages. The government does not acknowledge any oral claims for land ownership from the first inhabitants and only accepts the title deeds given by the Land Registry. This historical spatial division of former princely domains – supernatural places – along with Muslim and

Christian neighbourhoods, has also been overlaid with an administrative territorial division (Mearns 1999: 28–30). In short, in Ambon City both spatial boundaries and the boundaries of ethnic identity were complex and fluid.

The 1997 economic crisis in eastern Asia made the situation in and throughout the Moluccas tense. The demand for new jobs increased while the government, the main employer, became hesitant to fill any vacancies. At the same time the established route to office jobs – through patronage – was blocked, or at least constricted, because – after the fall of the long-standing Indonesian dictator, President Suharto, in May 1998 – there were attempts to prevent nepotism. The new route to the bureaucracy was to be through electoral success in the parliamentary elections of 1999, the first free elections in more than four decades. Local elites, who thought ethnicity would be instrumental in mobilizing a following, promoted the existing religious sentiments (Sidel 2001; Van Klinken 2001).

Against this backdrop, it is useful to turn to the national capital, Jakarta, where gangs of thugs dominated the streets. Most gangs were recruited on the basis of ethnicity and each gang occupied and defended their own turf (Melongok 1997). Muslim and Christian Ambonese gangs, along with many other gangs of varied ethnic composition, were hired by politicians in Jakarta in order to stage demonstrations or intimidate opponents. On 22 November 1998, fighting broke out between rival Ambonese gangs outside a gambling den in Jakarta. Rumours were spread that a mosque in a nearby Muslim neighbourhood had gone up in flames. The fights in Jakarta between the rival gangs escalated, resulting in several deaths and seven churches being burned to the ground. There were persistent but unproven rumours that the riots had been instigated on purpose and that the rival gangs were connected to one or the other of former president Suharto's two children, who were themselves rivals. The Jakarta city administration 'solved' this problem by expulsion, transferring dozens of Ambonese thugs from the capital back to Ambon (Aditjondro 2001).

The returning thugs accelerated competition in Ambon over resources, ranging from fluctuating boundaries of power to local protection rackets, at precisely the time that mobilization for the elections was under way. The possible decisive margin between Christian and Muslim voters made the political competition fierce. Fighting broke out on 19 January 1999 in Ambon City over a minor incident at a bus terminal, a public space that is divided in Indonesia between ethnic groups engaged in turf wars. News of the violence spread, getting coupled with rumours that led to a ripple of violence throughout Ambon, creating a climate of fear, mounting mutual suspicion and vengefulness. At first, the description of these skirmishes was limited to 'Ambonese Protestants fighting against the "Butonese" immigrants'. However, they quickly turned into a battle between Muslims and Protestants, regardless of place of origin (Sidel 2001; Van Klinken 2001).

Fire was a major weapon and literally hundreds of houses, mosques and churches were destroyed. Children aged between seven and twelve were given jerry cans full of petrol and induced to act as arson squads, backed by men armed with machetes and homemade guns. In July 1999, the fighting in Ambon City resumed. This time the Chinese-owned business district was also devastated, for reasons that remain rather unclear. As had been the case in January, each of the rival groups tried in vain to reach and destroy 'the heart' of the other's turf – that is, the Al Fatah mosque or the Maranatha church, both physically separated by just 300 metres. Catholics, who had until then stayed aloof, also became Muslim targets in this second riot.

Meanwhile, at the national level, both the Muslim and the secular (anti-Muslim) political parties pressed on with the exploitation of religious sentiments. The militant Muslim politicians in Jakarta called for a *jihad* (holy war fought by Muslims as a religious duty) in Ambon, blaming the present violence in the country on the Christian community. The *jihad* forces were connected to army units who were able to foresee that they would be called in to restore order. Christians mobilized similar support in Europe and North America through their church networks (Sidel 2001).

These manifestations of violence left Ambon City (and the Moluccas as a whole) more deeply segregated than it had ever been before. Muslim forces now occupy about 40 per cent of Ambon City and the Protestants 60 per cent. Barricades of oil drums mark the boundaries between them, criss-crossing the city. Subsequent fighting did not change the size of territories, but did eradicate enclaves. The city, in so far as it is not already razed to the ground, has been divided into two chunks: 'Muslim' and 'Protestant' territories. Even the state shipping line, which connects Ambon with the rest of Indonesia, got involved; it now either carries only Muslims or only Christians, after a series of furtive stabbings on board. With Muslim passengers the ship docks in the city harbour whereas, with Christians on board, the same ship moors at the naval base (Van Klinken 2001). It is estimated that between 4,000 to 10,000 people died in the fighting in Ambon and the rest of the Moluccas, in addition to the hundreds of thousands who were displaced (Aditjondro 2001: 100).

The outcome so far is that the once fluid, multiple and disunited spatial and ethnic boundaries have become fixed, one-dimensional and totally coextensive. One is either Muslim and living in the Muslim half of Ambon City or Christian and living in the Christian half. It is worth noting that the road dividing the two contesting sectors has been dubbed Jalan Gaza (Gaza Road). This partial misnomer, for in all likelihood the Ambonese have Jerusalem and not the Gaza Strip in mind, shows that the representation of an ethnic–spatial boundary from one place can also be used in another place. The symbolic meaning of Jerusalem is also well known in Ambon.

On the other cases presented in this volume

In the rest of this introduction, the remaining chapters will be briefly presented, beginning with the macro-, passing on to the meso- and finally looking at the micro-level of analysis of urban ethnic encounters. The cases in this book have been lumped together according to one of the three levels of analysis. This has been a difficult task since the authors have adopted perspectives at different levels and have integrated different approaches to ethnicity. We believe that such a combination of different levels of analysis and different approaches is invariably fruitful, creating a pool from which important conclusions can be drawn.

In Chapter 2, Joe Darden explicitly adopts the human ecology approach in his analysis of Toronto, Canada. He uses the 1996 census data to estimate the degree of residential segregation of Southeast Asians in Toronto. He provides a simple formula to calculate the degree of residential ethnic segregation, measured on a scale from 0 to 100. Though his formula requires a reliable census, even mathematical illiterates can easily understand it. The smaller the census tracts and the closer in size they are, the more precise is the outcome. At the top end of his scale there is total segregation, where ethnic groups do not mix at all. At the bottom of the scale, there is total integration and the ethnic composition in every census tract is equal to the ethnic composition of the city as a whole. Another methodological tool provided by Darden is a composite index for neighbourhood quality. The final step in Darden's analysis is comparing the findings of these two variables. He concludes that there is a high degree of dissimilarity, to the extent that more than half of the Southeast Asians in Toronto would have to change places of residence with whites – who form the majority population in Toronto – in order to become totally integrated. The neighbourhoods where Southeast Asians tend to cluster are, on average, of a lower socioeconomic quality than the white neighbourhoods.

In Chapter 3, another case from Toronto, Chandrakant Shah starts from the innovative and provocative premise that religious individuals or those who consider spirituality important have better quality of health and life expectancy. After providing ample evidence for this perhaps controversial premise, Shah continues his analysis by arguing that religious activity – in particular, attendance at a place of worship – has a beneficial impact on people's health. Implicit in his hypothesis is the assumption that ethnic groups will seek out houses of worship when taking up residence in a city or get organized to construct them. He elaborates this argument through a case study involving four religious minorities, with each religion associated with either one ethnic group or a combination of ethnic groups living in Toronto. Maps show the spatial distribution of places of worship and the respective followers of these denominations. In other words, certain ethnic groups are plotted on a map of Toronto together with their places of worship. These maps indicate a reciprocal relationship between the places of worship and the ethnic groups populating particular neighbourhoods. For instance, the Buddhist community is concentrated in the southern part of Toronto where there are also a fair number of Buddhist temples; by contrast, Buddhists in the northwest and east of the city are not so well served. One inference that can be made from Shah's hypothesis and analysis, with potentially wide-ranging consequences for public health policy, is that due to reduced access to these temples the Buddhists in the northwest and east of Toronto tend to be, on average, less healthy than the Buddhists in south

Toronto. Unfortunately, there is no data with which to substantiate or reject this hypothesis. This argument could *mutatis mutandis* be extended to every religious community in any city.

Religion is a concern of supreme importance in what is possibly the best known ethnically divided city in the world: Jerusalem. This holy city hosts people from three religions: Muslims, Jews and Christians. It is also a 'liberated' capital for the Israelis but an 'occupied' one for the Palestinians. Roadblocks here and there demarcate religious and ethnic divisions. Primordial sentiments run high. The question of what is to be done with Jerusalem is the most mind-boggling issue in the peace arrangements for the Palestinian–Israeli conflict, which is one of the most protracted political conflicts since World War II (Friedland and Hecht 1996). For most Arab and Jewish inhabitants of Israel, daily life involves the issues sketched by Thabet Abu Rass in Chapter 4, a crucial contribution to the theme of this volume. Abu Rass focuses on a number of small towns in Israel that together form an extended urban agglomeration known as the 'Little Triangle'. Fully in line with Ernest Burgess' model, the competition between the two contesting groups – the Arabs and the Israelis – has mainly been over the control of land. However, in contradiction of the basic assumption of Burgess, both groups value land not merely in economic terms but, first and foremost, as the basis of national existence. Moreover, state regulations have interfered in the land market. The state policy in Israel has contributed to the loss of land to the Arabs, as well as Jewish migrants invading formerly exclusively Arab land. The building of Jewish towns, the enlargement of existing Jewish towns and the construction of the Trans-Israel Highway have fragmented the originally uniform Arab region. Local Arab leaders feel that the development of the Arab towns has been restricted by the state policy of Israel.

Strategic considerations completely overruled the market for land in Beirut at the time of the civil war in Lebanon. Militias carved out sectors of the city in which to build their respective strongholds and forced an ethnic segregation upon the once prosperous city. Samir Khalaf dubbed this turf war 'the territorialization of identities' (Khalaf 1993: 28). In certain quarters, unearthing the slogans chalked on walls reveals the history of control of particular areas by successive militias. According to Daniel Genberg in Chapter 5, the inhabitants did not want their city to be divided into Muslim and Christian zones. Genberg gives an account of how people in Beirut experience and perceive the ethnically and religiously segregated urban space. Although the war has been over for a decade, on the mental map of the residents there are still places to be avoided or to which they simply never want to go. Part of the reconstruction work consists of creating neutral meeting places for Muslims and Christians. The fact that the city administration has left the urban planning to a French real estate company, SOLIDERE – an outsider unrelated to any of the former contestants – is of some significance. This chapter, apart from the fact that it deals with one of the political hotspots in the world, is highly relevant to our theme in that it provides a rare analysis of a process, however painful, of ethnic integration, starting from a state of total segregation.

In the colonial cities in Southeast Asia, Western capitalism existed side by side with a pre-industrial bazaar economy. According to T.G. McGee, the result was a mosaic consisting of the port, the western commercial district, a number of indigenous local markets scattered throughout the cities, several quarters dominated respectively by Chinese, Arab or Indian merchants, and indigenous neighbourhoods with a rural appearance. After Independence, the residential pattern based on ethnicity was replaced by a pattern based on social strata, because the elite of each ethnic group, including the Chinese, moved into the affluent neighbourhoods abandoned by Western colonials (McGee 1967: 126–41; see also Wertheim 1956). In Chapter 6, Harald Leisch takes McGee's well-known analysis further in his study on the Chinese in Jakarta, Indonesia. The spatial reflection of the special status granted to the Chinese in Jakarta is most visible in Glodok, the Chinese quarter of the old city, even though the Indonesian state has forbidden the display of any Chinese characters or any other exhibition of Chinese culture. During the last two decades, the post-war trend of the Chinese moving out to formerly elite colonial quarters has been reinforced by the development of luxurious new towns on the outskirts of the metropolitan region. These elite residential complexes have come to be predominantly, but not exclusively, inhabited by the Chinese. It is interesting that, unlike the immigrant Chinese architecture specific to Glodok and present in the idiosyncratic colonial atmosphere of the former elite neighbourhoods, the new towns do not exhibit any signs of Chinese identity. On the contrary, these residential complexes are under the strong influence of American architecture and design, as well as the American way of life more generally – including shopping malls and jogging tracks.

A remarkably similar spatial organization of isolated elite zones is found in a completely different setting – namely, the large Brazilian cities. As analysed by Carmen Rial and Miriam Grossi in Chapter 7, a discourse of violence in the public media has helped to create a 'culture of fear' amounting to paranoia. The response of the social elite to their own fears has been to claim part of the public space as private space and fence it off from the commoners. The elite live in guarded condominiums and drive in cars with smoked-glass windows, finding entertainment in guarded social clubs and shopping malls. The bipolar existence of the elite shuttling between residential areas and places of entertainment has created a mosaic pattern in Brazilian cities that consists of elite-class domains interspersed with lower-class zones. Despite the geographical proximity of the areas claimed respectively by the elite and the lower-class groups, the two worlds are almost completely separated, as if they were adversaries. The security industry has not only built a defence around elite residences, but has sometimes also constricted the lower-class neighbourhoods. For instance, one *favela* (squatter area) in Salvador is surrounded by a four-kilometre wall; most of its 10,000 residents have to make a long detour to go to other parts of the city. In reality, the level of violence in the lower-class neighbourhoods is actually much higher than in the elite neighbourhoods. Computer maps of the São Paulo police, with 'hot' and 'cold' zones, instantly register the location of each reported crime. The spots notorious for violence are 'no-go zones' even for the official enforcers of the law. Clearly, the clash of interests and mutual recriminations between social classes in Brazil carries ethnic undertones, because the upper-class/middle-class/working-class distinctions in the country roughly correspond with the white/mulatto/black continuum of the population. In this respect, the Brazilian case confirms the notion that ethnicity is constructed and not given or self-evident. The fact that, for instance, Japanese and Koreans – seen as non-white in northern America – are considered *branco* (white) in Brazil is a case in point.

At a meso-level of analysis, the chapters on Brazil and Jakarta focus on the level of the individual neighbourhood. Just as Wirth described in *The Ghetto* (1928) how a Jewish neighbourhood gradually developed in Chicago, Jian Guan depicts, in Chapter 8, how a Chinatown developed in Philadelphia, USA, during the course of the nineteenth century. The Chinese immigrants have settled in this city at the expense of the older settlers from Britain, Germany and Ireland. She documents how the new immigrants founded typically 'Chinese businesses', such as laundries and restaurants. At present, non-Chinese residents still account for only a quarter of the population of Chinatown. Nonetheless, a considerable number of the residents of Chinatown have moved out, so the majority of the Chinese in Philadelphia live elsewhere in the city. In Chinatown, bilingual street signs, phone booths decorated with pagodas, the Chinese character for long life (*shou*) on the sidewalks and other ethnic symbols add to the Chinese flavour. The value of this chapter lies in the fact that Guan shows the interplay between the neighbourhood, with its physical setting and network of social relationships, and the encroaching urban renewal projects, such as an expressway, a convention centre and a proposed stadium. Guan maintains that Philadelphia Chinatown has become a victim of ethnic insensitivity, since two blocks of housing were demolished without any regard to the needs of the community or its right to self-determination. The Chinese response has been an increased ethnic awareness on the part of the residents and a mobilization of Chinese support throughout the country. A giant mural depicts the history of the resistance of the residents of Chinatown.

In Chapter 9, Shenglin Chang presents a completely different type of Chinese neighbourhood, an affluent one, in a pioneering study that is an important contribution to the growing literature on Chinese settlements and the transnational way of life. Her chapter, in a way, reverses the topic of this volume. Instead of discussing how several ethnic groups live together in one spot, she analyses how one ethnic group lives in two different settlements. Taiwanese high-tech families show a pattern of circular, transnational migration, moving back and forth between Silicon Valley in California and Hsinchu Science-based Industrial Park (HSIP) in Taiwan. Bamboo Village – the residential section of HSIP favoured by these families, which displays typical American suburban standards (such as a low residential density, green spaces, homogeneity of social class and so on) – has distanced itself socially and economically from its surroundings. When the same Taiwanese elite live in California, they spend a considerable part of the time available for social activities at the *Ranch 99* megastores, a chain that offers services to predominantly Asian customers. This again isolates them to a certain degree from the other ethnic groups known to be existing in the vicinity. These high-tech immigrants rarely visit the Chinatown in San Francisco, which they consider a backward enclave of poor immigrants from the People's Republic of China. The result of this ongoing trans-Pacific movement between Taiwan and Silicon Valley has been a transcultural lifestyle with many common features.

In contrast to HSIP, which materialized according to an architectural blueprint, the neighbourhood in Tehran, Iran, described with a keen eye for detail by Soheila Shahshahani in Chapter 10, has developed gradually. Its crooked streets give away its rural origins. The neighbourhood is rich in history, starting with its name – derived from the neighbourhood's pioneer settler in the nineteenth century. Other families who once owned land there also have streets named after them. These peculiarities probably escaped the Kurd and Afghani newcomers. The various ethnic groups in the neighbourhood correlate, roughly, with particular socioeconomic levels. There is little inter-ethnic contact between the male residents, who mostly work in another zone of Tehran. Such contacts are reserved for special occasions and usually limited to the doorstep – when, for instance, practising the Islamic custom of taking over a dish of food on the occasion of a neighbour's bereavement. The womenfolk cross the ethnic boundaries more easily and visit each other, their kitchens being the centres of this conviviality. However, the children refrain from going to the houses of acquaintances and friends from other ethnic groups or classes, and prefer only to play together on a certain plot on a local boulevard. Hence, they somewhat reflect their parents' concern about their level of amicability with neighbours.

The example of the Tehran neighbourhood again shows that the borderlines between the macro-, meso- and micro-levels are blurred. At the micro-level, the scholar focuses on face-to-face contacts, which automatically restricts the study to limited spaces – that is, to a street or a square. Several chapters in this book are also partly devoted to a micro-level analysis, focusing on how and where people from different ethnic groups interact. For example, in Chapter 3, Chandrakant Shah describes how the construction of a Hindu temple met with opposition from non-Hindu neighbours and how a mosque in the outskirts of Toronto became the pivot of a newly mushrooming Islamic neighbourhood. In Chapter 10, people in a Tehran neighbourhood come together to rejoice in the streets during special occasions, such as after an important football victory. The local boulevard is also the place where the residents gather for an annual ritual feast, during which they jump over a fire. Likewise in Beirut (Chapter 5), the boulevard is a meeting place, where people of different denominations and from different ethnic groups love to stroll.

Chapter 11, by Jeffry Hester, focuses on one single shopping street in Osaka, Japan. The Koreans who sell ethnic Korean goods, such as ingredients for Korean dishes, own half of the shops in this street. When the shopping street began to suffer from increasing competition from the shopping malls nearby, the association of merchants started to think about ways to boost the appeal of the street to customers. One result has been the redecoration of the street with idiosyncratic Korean street lamps, coloured paving stones and archways at the entrance, changing it into a 'Korea Town'. By focusing his magnifying lens on a restricted space, Hester is able to show in detail how people were concerned about and behaved in relation to this street. But the special quality of this chapter lies in the fact that he does not lose sight of the developments in the wider, even national society. The Korean refurbishing of this particular shopping street was, in fact, the product of extended negotiations between shop owners, the local government and investors. The attraction of the idiosyncratic Korean design is a valuable marketing asset, perhaps also a means of production, for the retailers. At the same time, Korea Town is an object and milieu for consumption for the customers who come to shop, to hang around, to meet friends, to show off the fashion expressive of their personal lifestyles and so on. Korea Town has multiple symbolic meanings, depending on whose perspective is taken into account. One such meaning is the expression of a new and national civic ideology of multi-ethnic existence – 'living together' (with difference) – which is a substitute for the former and tacit expectation that foreigners should completely assimilate with Japanese society. Some Japanese shop owners in the street, however, oppose this new interpretation. Their opposition is visible in the inconsistency in design – due to their insistence on the Japanese way – of the street furniture used in Korea Town.

Chapter 12, by Heidi Dumreicher, analyses a street festival in a square in Vienna, Austria. She is able to provide a unique insight, owing to her being both observer and participant, taking roles not readily available to most anthropologists. That is, she is an organizer of the festival – instigated by the Oikodrom Forum, an NGO founded by herself – and, when she refers to a woman serving tea during the festival, that woman is actually herself. The aim of this activity, dubbed 'the Waterworld Festival', was to breathe life into a vast square that people usually wanted to leave behind them as quickly as possible. The festival deliberately strove not only to attract people, but also to bring different ethnic groups together. The square became a stage where, for example, a Turkish storyteller, young children jumping about on an inflated 'bouncy castle', breakdancing adolescents and others could perform or entertain side by side.

In contrast to the example from Vienna, in Chapter 13, Eveline Dürr analyses two cases of how different ethnic groups can be in competition over the control of public space in the old city centre of Albuquerque, USA. The main attraction of her study is that she shows how Anglo-Americans, Hispanics and American Indians not only contested the space physically, but also sought to represent their roots through it. In the first case she recounts, the itinerant Indian vendors were initially banned from the streets after a court case. Then the Anglo-American shop owners decided to have the American Indians back on the street, selling their ethnic goods from colourful blankets in front of their shops. The Indians had given Albuquerque a mystical and traditional appeal that attracted tourists. It is worth noting, though, that selling off blankets was not in fact the authentic Indian way of doing business, but a recent invention in keeping with the romanticized representation of the primitive but charming native. In the other case taken from Albuquerque, the Anglo-Americans were the adversaries of the Hispanics. The conflict between them was about whether the city centre had a predominantly economic function, as claimed by the Anglo-American shop owners, or a cultural and religious meaning, as the Hispanics believed it did.

With this last chapter, the trip around the globe comes to an end. The variety in the content of the chapters, ranging from the macro- to the micro-level of analysis and covering three different continents, provides a unique opportunity for cross-cultural comparative analysis across the globe. The comparison between these cases enables us, to a certain extent, to probe answers to the questions of how inter-ethnic relations can create and pattern urban space and how inter-ethnic relations are organized in urban spaces. Such an examination invariably results in a number of theoretical observations. They are presented in our conclusion (Chapter 14), where we refer also to other pertinent studies in order to expand the empirical basis of the comparative analysis. Taking into account the limited number of cases presented here, it is clear that the theoretical remarks do not claim general validity. They must be considered as hypotheses to be elaborated and refined through further research.

References

Aditjondro, George Junus (2001) 'Guns, pamphlets and handie-talkies: How the military exploited local ethno-religious tensions in Maluku to preserve their political and economic privileges', in Ingrid Wessel and Georgia Wimhöfer (eds), *Violence in Indonesia*, Hamburg: Abera.

Anderson, Benedict R. O'G. (1992) *Long-Distance Nationalism: World capitalism and the rise of identity politics* (the Wertheim Lecture 1992), Amsterdam: CASA.

Appadurai, Arjun (1996a) 'Disjuncture and difference in the global cultural economy', in Arjun Appadurai, *Modernity at Large*, Minneapolis and London: University of Minnesota Press.

—— (1996b) 'Life after primordialism', in Arjun Appadurai, *Modernity at Large*, Minneapolis and London: University of Minnesota Press.

Appleyard, Donald (with Gerson, M. Sue and Lintell, Mark) (1981) *Livable Streets*, Berkeley, Los Angeles and London: University of California Press.

Banks, Marcus (1996) *Ethnicity: Anthropological constructions*, London: Routledge.

Barth, Fredrik (1969) 'Introduction', in Fredrik Barth (ed.), *Ethnic Groups and Boundaries: The social organization of cultural difference*, Boston: Little, Brown and Company.

—— (1994) 'Enduring and emerging issues in the analysis of ethnicity', in Hans Vermeulen and Cora Govers (eds), *The Anthropology of Ethnicity: Beyond 'Ethnic Groups and Boundaries'*, Amsterdam: Het Spinhuis.

Boyer, Christine (1983) *Dreaming the Rational City: The myth of American city planning*, Cambridge: MIT Press.

Brass, Paul R. (1991) *Ethnicity and Nationalism: Theory and comparison*, New Delhi, Newbury Park and London: Sage.
Burgess, Ernest W. (1967, orig. 1925) 'The growth of the city: an introduction to a research project', in Robert E. Park, Ernest W. Burgess and Roderick D. McKenzie, *The City*, Chicago: University of Chicago Press.
Castells, Manuel (1975) *La question urbaine* (The Urban Question) (revised edn), Paris: Maspero.
—— (1996) *The Information Age: Economy, society and culture*, vol. 1, *The Rise of the Network Society*, Cambridge: Blackwell.
Castels, Stephen (2000) *Ethnicity and Globalization: From migrant worker to transnational citizen*, London, Thousand Oaks and New Delhi: Sage.
Cohen, Abner (1974) 'Introduction: The lesson of ethnicity', in A. Cohen (ed.), *Urban Ethnicity*, London: Tavistock.
Colombijn, Freek (1992) 'Dynamics and dynamite: Minangkabau urban landownership in the 1990s', *Bijdragen tot de Taal-, Land- en Volkenkunde* 148: 428–64.
—— (1994) *Patches of Padang: The history of an Indonesian town in the twentieth century and the use of urban space*, Leiden: Research School CNWS.
Cressey, Paul G. (1932) *The Taxi-Dance Hall: A sociological study in commercialized recreation and city life*, Chicago: University of Chicago Press.
Dekker, T., Helsloot, J. and Wijers, C. (eds) (2000) *The Construction of Ethnic Identities*, Amsterdam: Het Spinhuis.
de Vos, George A. (1996) 'Ethnic pluralism: conflict and accommodation; the role of ethnicity in social history', in Lola Romannucci-Ross and George de Vos (eds), *Ethnic Identity: Creation, conflict, and accommodation*, Walnut Creek: Alta Mira Press.
Eriksen, Thomas Hylland (1993) *Ethnicity and Nationalism: Anthropological perspectives*, London and East Haven: Pluto.
Evers, Hans-Dieter and Schiel, Tilman (1988) *Strategische Gruppen: Vergleichende Studien zu Staat, Bürokratie und Klassenbildung in der dritten Welt* (Strategic Groups: Comparative studies on state, bureaucracy and class formation in the Third World), Berlin: Dietrich Reimer.
Friedland, Roger and Hecht, Richard (1996) *To Rule Jerusalem*, Cambridge: Cambridge University Press.
Gans, H.J. (1982, 1st edn 1962) *The Urban Villagers: Group and class in the life of Italian-Americans*, New York: Free Press.
Geertz, Clifford (1973) 'The integrative revolution: primordialist sentiments and civil politics in the new State', in Clifford Geertz, *The Interpretation of Cultures: Selected essays*, New York: Basic Books.
Goldberg, Michael and Chinloy, Peter (1984) *Urban Land Economics*, New York: Wiley.
Gottdiener, Mark (1985) *The Social Production of Urban Space*, Austin: University of Texas Press.
Gottdiener, Mark and Feagin, Joe R. (1988) 'The paradigm shift in urban sociology', *Urban Affairs Quarterly* 24: 163–87.
Gould, Peter and White, Rodney (1992, 1st edn 1974) *Mental Maps*, London: Routledge.
Guldin, Greg (1985) 'Measuring urban ethnicity', in Aidan Southall, Peter J.M. Nas and Ghaus Ansari (eds), *City and Society: Studies in urban ethnicity, life-style and class*, Leiden: Institute of Cultural and Social Studies, University of Leiden.
Hamnett, Chris (2001) 'Social segregation and social polarization', in Ronan Paddison (ed.), *Handbook of Urban Studies*, London, Thousand Oaks and New Delhi: Sage.

Hannerz, Ulf (1980) *Exploring the City: Inquiries toward an urban anthropology*, New York: Columbia University Press.
—— (1996) *Transnational Connections: Culture, people, places*, London: Routledge.
Harvey, David (1990) *The Condition of Postmodernity: An enquiry into the origins of cultural change*, Cambridge and Oxford: Blackwell.
Jacobs, Jane (1964, orig. 1961) *The Death and Life of Great American Cities*, Harmondsworth: Penguin.
Jenkins, Richard (1996) *Social Identity*, London and New York: Routledge.
—— (1997) *Rethinking Ethnicity: Arguments and explorations*, London: Sage.
Khalaf, Samir (1993) 'Urban design and the recovery of Beirut', in Samir Khalaf and Philip S. Khoury (eds), *Recovering Beirut: Urban design and post-war reconstruction*, Leiden: Brill.
Kilian, Ted (1998) 'Public and private, power and space', in Andrew Light and Jonathan M. Smith (eds), *The Production of Public Space*, Lanham: Rowman & Littlefield.
Lefebvre, Henri (1986, orig. 1974) *La production de l'espace* (The Production of Space), 3rd edn, Paris: Éditions anthropos.
Lo, Fu-chen and Yeung, Yue-man (eds) (1996) *Emerging World Cities in Pacific Asia*, Tokyo et al.: United Nations University Press.
Lynch, Kevin (1960) *The Image of the City*, Cambridge: MIT Press.
McCarthy, John (1999) 'Chicago: a case study of social exclusion and city regeneration', *Cities* 16: 323–31.
McGee, Terry G. (1967) *The Southeast Asian City: A social geography of the primate cities of Southeast Asia*, London: Bell.
McGee, Terry G. and Robinson, Ira M. (eds) (1995) *The Mega-Urban Regions of Southeast Asia*, Vancouver: UBC Press.
Mearns, David (1999) 'Urban kampongs in Ambon: Whose domain? Whose desa?', *The Australian Journal of Anthropology* 10(1): 15–33.
Melongok (1997) 'Melongok kantong-kantong preman ibu kota' (Inspecting the Pockets of Thugs), *Forum keadilan* 6(17): 106–7.
Nas, Peter J.M. and Houweling, Antonia J. (1998) 'The network metaphor: an assessment of Castell's network society paradigm', *Journal of Social Sciences* 2: 221–32.
Oberschall, Anthony (2000) 'The manipulation of ethnicity: From ethnic cooperation to violence and war in Yugoslavia', *Ethnic and Racial Studies* 23: 982–1001.
Öncü, Ayşe and Weyland, Petra (1997) 'Introduction: struggles over *lebensraum* and social identity in globalizing cities', in Ayşe Öncü and Petra Weyland (eds), *Space, Culture and Power: New identities in globalizing cities*, London and New Jersey: Zed Books.
Paddison, Ronan (ed.) (2001) *Handbook of Urban Studies*, London, Thousand Oaks and New Delhi: Sage.
Park, Robert E. (1967, orig. 1925) 'The city: suggestions for the investigation of human behavior in the urban environment', in Robert E. Park, Ernest W. Burgess and Roderick D. McKenzie, *The City*, Chicago: University of Chicago Press.
Park, Robert E., Burgess, Ernest W. and McKenzie, Roderick D. (1967, orig. 1925) *The City*, Chicago: University of Chicago Press.
Phillips, E. Barbara (1996) *City Lights: Urban–suburban life in the global society*, 2nd edn (1st edn 1981, E. Barbara Phillips and Richard T. LeGates), New York and Oxford: Oxford University Press.

Rogers, Alisdair and Vertovec, Steven (1995) 'Introduction', in Alisdair Rogers and Steven Vertovec (eds), *The Urban Context: Ethnicity, social networks and situational analysis*, Oxford and Washington: Berg.
Sassen, Saskia (1991) *The Global City: New York, London and Tokyo*, Princeton: Princeton University Press.
Schefold, Reimar (1998) 'The domestication of culture: nation-building and ethnic diversity in Indonesia', *Bijdragen tot de Taal-, Land- en Volkenkunde* 154: 259–80.
Sidel, John T. (2001) 'Two's company, three's an unruly crowd? Bossism and political violence in the Philippines, Thailand and Indonesia', paper presented at the 3rd EUROSEAS Conference in London, 6–8 September.
Tajbakhsh, Kian (2001) *The Promise of the City: Space, identity, and politics in contemporary social thought*, Berkeley, Los Angeles and London: University of California Press.
Van Klinken, Gerry (2001) 'The Maluku wars: bringing society back in', *Indonesia* 71: 1–26.
Verdery, Katherine (1994) 'Ethnicity, nationalism and state-making', in Hans Vermeulen and Cora Govers (eds), *The Anthropology of Ethnicity: Beyond 'Ethnic Groups and Boundaries'*, Amsterdam: Het Spinhuis.
Wertheim, Willem Frederik (1956) 'Urban development', in Willem Frederik Wertheim, *Indonesian Society in Transition: A study of social change*, The Hague and Bandung: Van Hoeve.
Wirth, Louis (1928) *The Ghetto*, Chicago: University of Chicago Press.
—— (1938) 'Urbanism as a way of life', *The American Journal of Sociology* 44: 1–24.

Part I
The macro-level analysis of urban ethnic encounters
Enclaves and the zones of the cities

2 Residential segregation and neighbourhood socioeconomic inequality
Southeast Asians in Toronto

Joe T. Darden

Theoretical framework

This research, although influenced by human ecological models related to immigrant settlement in the United States which described the spatial assimilation of white immigrant groups from Europe (Alba and Logan 1991, 1993; Massey 1985), approaches the issue from a perspective that is also different from most previous and similar research on Canada. It assumes that in Canada, as in other white-dominated industrial societies, visible minorities (that is, people of colour) are set apart from the white majority. They are 'racialized' through practices and/or policies imposed by the white majority. These 'racialized' minorities have become an increasing concern since Canada changed its racially restrictive (whites preferred) immigration policy in 1967. Since then, immigrants have become qualified not on the basis of their country of origin or ethnic background as was the case in the past, but based on points given for education qualifications, job experience and other criteria (Department of Manpower and Immigration 1974: 27–34; Henry 1994).

These changes in policy resulted in a steady flow of immigrants of colour (mostly from Asia, the Caribbean, Latin America and Africa). Today, more talented immigrants from non-European countries are arriving in Canada. They are transforming Canada's racial composition from a largely white society to a racially diverse one. In 1996, Statistics Canada defined about 3.2 million people in Canada's population as non-white visible minorities, representing 11.2 per cent of the total population (Statistics Canada 1997). Of these visible minorities, 70 per cent were immigrants and almost all (94 per cent) lived in large metropolitan areas, most of them in Toronto (Statistics Canada 1998). However, there is a lack of empirical assessment of how those new arriving minorities have been incorporated into Canadian society. Such an assessment is crucial because most immigrants of colour were allowed into Canada due to a labour shortage that required a change in the country's race-restrictive immigration policy. Since the change in policy was motivated by labour needs instead of a change in white behaviour (namely, associated with a more welcome attitude towards people of colour), incorporation of visible minorities into Canadian society is likely to be more difficult.

While most research on immigrant groups has assumed that social and economic equality will occur through spatial assimilation with the majority group, this model may not apply to non-whites in predominantly white societies. Some research in the United States has already revealed that the spatial assimilation model is less effective in predicting the housing and locational outcomes of blacks, Puerto Ricans and non-white Hispanics (Massey and Denton 1993). The model may especially not apply in a country such as Canada, where the policy towards immigrants is multiculturalism rather than spatial assimilation.

Multiculturalism

In 1971, Prime Minister Trudeau made a statement which became the ideological basis of a multiculturalism policy for Canada. Among the objectives of the 1971 policy were that (1) the government would support all of Canada's cultures and seek to assist, resources permitting, the development of those cultural groups that had demonstrated a desire to continue to develop and (2) the government would assist members of all cultural groups to overcome cultural barriers to full participation in Canadian society (*House of Commons Debates* 1971: 8581). Groups could apply for governmental financial assistance to carry out various cultural activities.

Due to criticism that the multiculturalism policy of 1971 did not pay enough attention to the rights of racial minorities, a Multiculturalism Act was passed in 1988. This act stated that 'multiculturalism' was a federal policy with the objectives of:

1 promoting full and equitable participation of individuals and communities of all origins,
2 enhancing the development of communities sharing a common origin,
3 ensuring equal treatment and protection for all individuals while respecting their diversity, and
4 promoting and recognizing multiculturalism as a fundamental characteristic of Canada.

(Statutes of Canada 1988)

In other words, the 1988 Act involved the federal government in promotion of both cultural retention and social equality (that is, equal access and participation for all Canadians in the economic, social, cultural and political life of the nation). But can such objectives be achieved in a nation that has a history of white supremacy?

White supremacy is an ideology which holds that, in any relations involving people of colour and whites, the white race must have the superior position (Rose *et al.* 1969: 68). White supremacy differs from other majority group supremacies. It has been used to exploit and control people of colour. According to Cornell West, white supremacy is the most vicious ideological articulation used in defence of power and privilege (West 2000: 272). It concerns more than just race *per se*, because it is a way of acknowledging the degree to which race is inextricably linked to asymmetric relations of power. In multiracial white-dominated societies where white supremacy is a prevailing value (Rose *et al.* 1969: 69; Smith 1989: 4), the white majority group is likely to restrict the opportunities for non-white groups to attain neighbourhood qualities equal to theirs (Alba and Logan 1993). The constraints imposed will vary depending on the reaction of the white majority towards each visible minority group (Darden 1989). The results will be manifested by the degree of residential segregation and neighbourhood racial inequality.

The consequences of residential segregation on social inequality

Social scientists have noted that the spatial organization of society, and *where* a group lives within that society, plays a role in shaping a group's social mobility and opportunities in that society (Blau 1977). The way a society is organized influences individuals and families, yet it lies beyond their individual control.

In a predominantly white, free-market society, where the prevailing ideology is white supremacy, opportunities, resources and benefits are not distributed evenly across neighbourhoods. Instead, certain neighbourhoods, usually predominately white ones, are given more prestige by the white majority. Such a prestige is complemented by the best schools, recreation facilities, highest house prices, safest streets and other neighbourhood amenities. In market-oriented, predominantly white societies, whether by perception or other means, metropolitan areas are spatially differentiated by neighbourhood and socioeconomic status that also links social mobility to spatial mobility. As groups attempt to get ahead, moving up is linked to moving out spatially. For disadvantaged groups – disproportionately made up of people of colour – the process has been called 'the geography of opportunity' (Rosenbaum 1995). It has resulted in many benefits, including access to better schools and jobs, reduced fear of crime and greater neighbourhood satisfaction (Briggs, Darden and Aidala 1999). Thus there is an incentive for the disadvantaged visible minority group to 'break out of its spatial isolation'.

Prior research on racial residential segregation and neighbourhood inequality in Canada

Unlike other multiracial white-dominated societies, such as the United States and Britain, Canada has played down the importance of race as a factor in residential segregation and neighbourhood inequality. Canada has chosen to emphasize ethnicity or ethnic origin (Balakrishnan 1976; Darroch and Marston 1971). Instead of comparing the white majority with blacks, researchers have compared Italians, British, Germans and Greeks, for example, with Jamaicans. Any inequality between the groups has been attributed to internal cultural characteristics. They often ignore the fact that Jamaicans' lower status when compared with white ethnic groups may be due to external forces such as racial discrimination. These studies have viewed ethnic residential segregation as a factor in identifying the extent to which ethnic groups share similar neighbourhood qualities with average Canadians. Canadian studies have found that levels of residential segregation, however, vary substantially even after controlling for periods of immigration and levels of socioeconomic resources. British and Northern European immigrants consistently experience low levels of segregation compared with other groups (Balakrishnan and Selvanathan 1990).

Jews, however, are highly segregated in Toronto, even though they live in neighbourhoods with the most desirable neighbourhood characteristics (Darden and Kamel 1999). Indeed, Darroch and Marston's study concluded that the residential location of the Jewish population in Toronto was least affected by socioeconomic differentials (Darroch and Marston 1971). Cultural and religious reasons probably best explain the location and segregation of the Jews. The Jewish example in Toronto is different from that of other groups and lends some evidence in support of the hypothesis that a white ethnic group can remain highly residentially segregated while retaining the most desirable neighbourhood characteristics in the highest quality neighbourhoods. We do not know whether the same experience applies to any visible minority group.

In sum, research on ethnic groups in Canada reveals an inconsistent relationship between ethnic group residential segregation, socioeconomic status and quality of neighbourhood. For some ethnic groups, a linkage exists, but for other groups it does not (Darroch and Marston 1971).

As for the small amount of research on racial residential segregation and neighbourhood inequality in Canada, the studies that are most closely related to the research in this paper have been done by Fong and Gulia (1996). Using 1991 census data by census tract and specially requested tables from Statistics Canada, Fong and Gulia compared the neighbourhood qualities of immigrants of various racial and ethnic groups in Toronto and Vancouver, two of Canada's largest metropolises. Using regression analysis, the authors concluded that, for British immigrants, the attainment of some neighbourhood qualities is affected by their education and income. For Chinese, the analysis suggests that the attainment of most neighbourhood qualities is not consistently affected by socioeconomic variables. Finally, for blacks, neighbourhood attainment is not related to their education, their income or their proportion in the tract. Results showed no consistent statistically significant effect of any factor on attainment of neighbourhood qualities for the blacks. The authors suggested that blacks may face problems of translating their socioeconomic status into desirable neighbourhood qualities (Fong and Gulia 1996: 141).

Objectives

This paper will focus on the extent to which Southeast Asians in Toronto have attained neighbourhood socioeconomic equality with white Canadians. The objectives, which are influenced by ecological theory, are threefold:

1   to determine the level of residential segregation between Southeast Asians and the white majority in the Toronto Census Metropolitan Area (CMA);
2   to determine the spatial distribution pattern of Southeast Asians and the white majority at the census tract or neighbourhood level; and
3   to determine the differential characteristics of neighbourhoods where the white majority and Southeast Asians reside.

Given the ideology of white supremacy prevalent in Canada – a multiracial, white-dominated society – it is hypothesized that, despite Canada's multiculturalism policy, Southeast Asians are experiencing both a high level of residential segregation from white Canadians and a high degree of inequality in socioeconomic neighbourhood characteristics. Thus they are not experiencing equal access and participation in the economic and social life of the Toronto Census Metropolitan Area (CMA). A CMA is a very large urban area (known as the 'urban core'), together with adjacent urban and rural areas (known as the 'urban and rural fringes') that have a high degree of social and economic integration with the urban core. A CMA has an urban core population of at least 100,000 based on the previous census (Statistics Canada 1997: 181).

Data and methodology

The Canadian 1996 Census Profile Series (Statistics Canada 1998) was utilized. The 1996 census was the first time a population group question was asked in the census. Respondents were presented with ten mark-in categories and one write-in box, and asked to mark or specify one or more of the following: white, Chinese, South Asian, black, Arab/West Asian, Filipino, Southeast Asian, Latin American, Japanese, Korean, or other (specify). Unlike the previous questions on ethnic origin, this question allows a *direct* analysis of residential segregation of *racial* groups. The population group question provides information for visible minority groups in Toronto and the rest of Canada – information that is required for programmes under the Employment Equity Act. According to this Act, members of a visible minority are people who are non-Caucasian in race or non-white in colour.

This research is the first opportunity to study racial residential segregation and neighbourhood racial inequality. The group selected for analysis is Southeast Asians. There were 46,240 Southeast Asians in the Toronto Census Metropolitan Area (CMA) in 1996, representing slightly more than 1 per cent of the total CMA population.

The 1996 census also provides most of the census socioeconomic variables for low levels of geography, including census tracts.[1] More specifically, social and economic variables from the Census Profile Series were used to construct a Composite Socioeconomic Index for Toronto CMA's census tracts to measure neighbourhood socioeconomic inequality.

Residential segregation index

The first objective of the study was attained by computing the level of residential segregation between whites and Southeast Asians in the Toronto CMA. Residential segregation is defined as the overall unevenness in the spatial distribution of two groups. The endeavour employs the index of dissimilarity D that measures residential segregation at the census tract level. Mathematically, the index of dissimilarity is stated as:

$$D = 100 \left(\frac{1}{2} \sum_{i=1}^{k} |x_i - y_i| \right)$$

where

K = Number of census tracts.

x_i = The percentage of the Toronto Census Metropolitan Area's Southeast Asian population living in a given census tract.

y_i = The percentage of the Toronto CMA's white population living in the same census tract i.

D = The index of dissimilarity. It is equal to half the sum of the absolute differences (positive and negative) between the percentage distribution of the white population and the Southeast Asian minority group population in the Toronto CMA (Darden and Tabachneck 1980; Duncan and Duncan 1955).

The value for the index of dissimilarity may range from '0' indicating no segregation to '100' indicating total segregation. This value stands for the minimum percentage of the population from either group that would have to move from one tract to another to achieve an even spatial distribution throughout the Toronto CMA. Despite its widespread use in residential segregation research, however, the index of dissimilarity has a certain limitation. The limitation is related to the fact that the index is sensitive to the size of the spatial units used, the smaller the spatial unit, the bigger the D (Darden and Haney 1978: 124). Since census tracts were chosen as the spatial units for this study, the degree and magnitude of segregation by block or enumeration area could not be determined. An enumeration area is the geographic area canvassed by one census representative. It is the smallest standard geographic area for which census data are reported (Statistics Canada 1997: 208). The segregation index in this study takes into account only differences in the spatial distribution of Southeast Asians and whites between census tracts, revealing nothing about the distribution of the same groups between blocks or enumeration areas.

A Composite Socioeconomic Index for Toronto CMA's census tracts

Difference in neighbourhood quality of life between whites and Southeast Asians was assessed using a Composite Socioeconomic Index (CSI). Eight socioeconomic status variables were used to calculate the CSI:

1   the percentage of university degrees,[2]
2   median family income,
3   percentage of managerial and professional status positions,[3]
4   average value of dwelling,
5   average gross rent of dwelling,
6   percentage of homeownership,
7   incidence of low income[4] and
8   unemployment rate.

Variables representing 'incidence of low income' and 'unemployment rates' were given a negative sign when calculating the index. This would allow the CSI to capture the depreciating effect of these variables on the quality of life in neighbourhoods and to provide an adequate coverage of each group's social and economic conditions. Overall, the eight socioeconomic variables allowed the study to capture the most important aspects of neighbourhood characteristics.

The eight variables characterize census tracts, which are surrogates for neighbourhoods. The variables are not the characteristics of individual white and Southeast Asian households. Operationalizing the concept of a neighbourhood by census tract is quite common among urban social scientists. As used here, the census tracts reveal the characteristics of the total population and housing located within them.

The assumptions behind the eight variables are that the most desirable neighbourhoods (census tracts) are those with:

1   the highest percentage of residents with university degrees,
2   the highest median family income and
3   the highest percentage of residents in managerial and professional status positions.

Such residents reside in homes with the highest average value or live in apartments with the highest average rent. Such neighbourhoods also have the highest percentage of homes that are owned by the residents. The least desirable neighbourhoods are those with the lowest incomes and the highest unemployment rates. Why are the neighbourhoods as identified by variables 1 to 6 the most desirable? This can be demonstrated by discussing variable 6 (homeownership).

Ownership has been favoured by Canadians because 'it provides the consumer with control' (Hannley 1993: 210). In Canada, like other predominately white societies, homeownership provides owners with a stake in the system. It is also perceived as an 'indicator of social status and a source of personal autonomy' (Agnew 1981: 75). Homeownership is 'an established path to status and security' and represents permanency and stability in life (Ray and Moore 1991: 2). Buying a home is an 'influential statement of success, security and stability' and a means of fitting into the social fabric (Adams 1984: 524).

In addition to the social advantages, the economic benefits of homeownership are also unquestioned. According to Saunders, homeowners at all class levels generally can and do profit from homeownership (Saunders 1978: 234). They do so because of house price inflation over time, trading up and declining housing costs as the mortgage is paid off and ownership becomes 'free and clear'. It can also be argued that homeownership provides the type of profits that cannot be achieved through most other market mechanisms by people of modest means (Verberg 2000: 171). Other researchers have also documented positive outcomes of homeownership, ranging from financial well-being to increased social status and personal security (Adams 1984; Agnew 1981; Perin 1977; Rakoff 1977; Sullivan 1989).

Homeownership is also positively related to political benefits in the form of political participation. It is suggested that homeowners are more likely than tenants to participate in mainstream political activities because the former category has a stake in the social and economic benefits of property ownership (Verberg 2000: 170). In one sense this is what Engels (1936) meant by political incorporation through homeownership. There is some evidence that high homeownership rates are associated with higher levels of voter turnout in Canadian elections (Pratt 1987).

In contrast, being a tenant is fundamentally different both socially and economically (Blum and Kingston 1984). Compared to homeowners, tenants are more likely to be viewed as an unsettling 'out-group' with a lack of social esteem (Agnew 1981: 75), besides the possibility of being seen as a transient group dependent on their landlords rather than an integrated part of society (Balakrishnan and Wu 1992). Economically speaking, federal, provincial and municipal policies are less likely to favour tenants and more likely to favour homeowners (Backer 1993). Tenants are penalized by the property tax treatment of apartments as 'commercial property', which makes rent increasingly expensive (Skaburskis 1996).

In sum, the view that homeownership provides social, economic and political advantages over renting for Southeast Asians is fairly well supported and should not be disputed. Thus neighbourhoods with high homeownership rates – or scoring highly on the other socioeconomic variables (1–5 and 7–8) – provide strong incentives for Southeast Asians to want to locate there. However, since these areas are disproportionately occupied by whites, racial residential integration is the only means to gain access to such neighbourhoods. Thus it is not necessarily the desire to integrate with whites that is primary, but the desire to improve one's socioeconomic quality of life. However, it is often difficult to achieve the latter without the former.

On the other hand, it is conceivable that some Southeast Asians might prefer residential segregation from whites. They would thus remain in areas mostly occupied by other members of their group, even if this meant a continuous lack of social mobility and poor quality of life through residence in low-income areas and areas of high unemployment. It is beyond the scope of this study to determine what percentage of Southeast Asians would make that choice.

I turn now to a discussion of how the analysis in this study was done. The Composite Socioeconomic Index (CSI) is to be stated mathematically as:

$$CSI_i = (\Sigma \frac{V_1}{V_{1CMA}} + \frac{V_2}{V_{2CMA}} + \frac{V_{n+1}}{V_{n+1CMA}})$$

where

V = Socioeconomic status variable for a given census tract i.
V_{CMA} = Socioeconomic status variable for Toronto CMA;
CSI_i = The Composite Socioeconomic Index for census tract i is the sum of socioeconomic status variables' weight relative to Toronto CMA's socioeconomic status.

Based on the formula, the index represents the sum of socioeconomic scores given to each of the census tracts in Toronto CMA. These scores were obtained through adding up the weight of each socioeconomic variable for each tract. Weighting is equal to the ratio for each social and economic variable per tract to the average value of Toronto CMA as a whole. In other words, each weight was calculated by the simple ratio method of dividing each social or economic characteristic for each tract by the average value of that socioeconomic characteristic for Toronto CMA. For example, a tract with family median income equal to $100,000 would have a weight of 2.0 if the average family median income for Toronto CMA is $50,000.

The CSI is then sorted to rank the city's neighbourhoods according to their quality of life. Thus Toronto CMA can be divided into four equal socioeconomic ranges with boundaries at the 25th, 50th and 75th percentiles of the CSI frequency distribution. These percentiles refer to neighbourhoods (census tracts) and not the population. Grouping based on four equal socioeconomic ranges allowed the division of neighbourhoods into four types with equal variation in each socioeconomic status (that is, upper, upper-middle, lower-middle and lower class – see Table 2.1).

Southeast Asians: characteristics

Most Southeast Asians (72 per cent) are immigrants from Vietnam. In addition to the Vietnamese, Statistics Canada considers the following to be Southeast Asians: Burmese, Cambodian, Laotian and Thai (Statistics Canada 1997: 16; Statistics Canada 1998). Most are refugees and have arrived relatively recently. Despite their recent arrival, however, most are Canadian citizens. Moreover, whatever the period of immigration, immigrants from Vietnam are considerably more likely than their counterparts to have Canadian citizenship (Minister Supply and Services Canada 1996: 4).

Immigrants from Vietnam are more highly urbanized than the Canadian-born population and more of them live in the Toronto CMA than in any other metropolitan area. Although the majority of immigrants from Vietnam can speak English or French, the two most commonly spoken languages at home are Vietnamese and Chinese. The largest share of immigrants from Vietnam are Buddhists. However, a large proportion of Vietnamese immigrants (more than a third) have no religious affiliation (Minister Supply and Services Canada 1996: 5).

Vietnamese immigrants are less likely than people in other groups to have a university degree, and are less likely to have a degree compared to the Canadian-born adult population. Instead, Vietnamese immigrants are more likely than all immigrants or people born in Canada to have less than a high-school education. Since most arrived in Canada as refugees, education was less of a criterion for entry (Hernandez 1994: 54). On the other hand, Vietnamese with post-secondary degrees were more likely than other immigrants and the Canadian-born population to be graduates of professional programmes in mathematics, engineering, applied science and the health professions.

Whether employment is with a firm or self-employment, Vietnamese immigrants are less likely than most immigrants or people born in Canada to be working. Vietnamese immigrants are also less likely than most workers to be employed full-time year round. Vietnamese immigrants have a relatively high unemployment rate.

When employed, Vietnamese immigrants are more likely than their counterparts in the total immigrant population and the Canadian-born population to work in menial jobs (45 per cent) (Statistics Canada 1998).

In general, the incomes of immigrants from Vietnam are lower than those of people born in Canada and lower than those of most other immigrants. Moreover, a relatively large proportion (41 per cent) of immigrants from Vietnam have incomes below Statistics Canada low-income cut-offs. Given the demographic characteristics described above, there are several key questions: Where do Southeast Asians live in the Toronto CMA? Do they share neighbourhoods with whites? If so, to what extent?

Spatial distribution of Southeast Asians and whites in Toronto CMA

A basic question was where the two groups lived in 1996. This was done by mapping the distribution of each group in each of Toronto CMA's census tracts relative to its average distribution across Toronto CMA. The distribution of each population, whites and Southeast Asians, was divided into two segments. The first segment contains census tracts where the white or Southeast Asian population is less than each group's average population in the CMA as a whole (that is,

Segregation and neighbourhood inequality   37

less than 1.09 per cent for Southeast Asians and less than 68.28 per cent for whites). The second segment, on the other hand, contains census tracts where the population is equal to or more than the average percentage population for each group in the CMA as a whole (that is, more than or equal to 1.09 per cent for Southeast Asians and more than or equal to 68.28 per cent for whites).

The results of this analysis are presented in Maps 2.1 and 2.2. The first map shows the spatial distribution of Southeast Asians. The population is relatively

Distribution of Southeast Asians in Toronto CMA
☐ Census tracts with zero Southeast Asians
▨ Census tracts with less than average Southeast Asians (< 1.09 %)
■ Census tracts with more than average Southeast Asians (> 1.09 %)

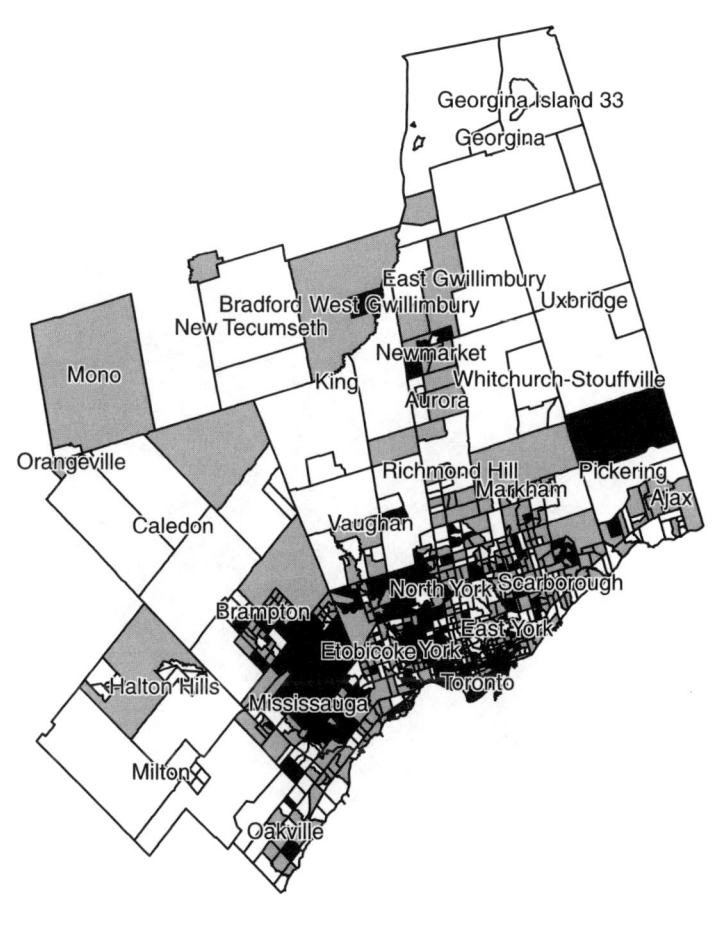

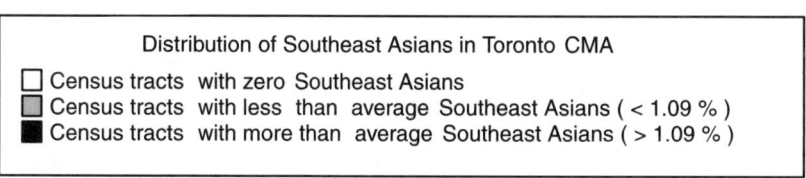

Map 2.1 Distribution of Southeast Asians across Toronto CMA's census tracts, 1996

dispersed in the southern half of Toronto CMA, with a few contiguous concentrations in Bradford, West Gwillimbury and Newmarket in the northern half. The population of neighbourhoods in Toronto CMA ranged from 0 to 16.74 per cent Southeast Asians. Spatial dispersion is more common when the group being examined is very diverse, as is the case for Southeast Asians. Map 2.2 shows the spatial distribution of the white population. Whites were found to follow a different spatial distribution pattern than Southeast Asians. The level of residential segregation that this distribution pattern reflects is 57 per cent. That

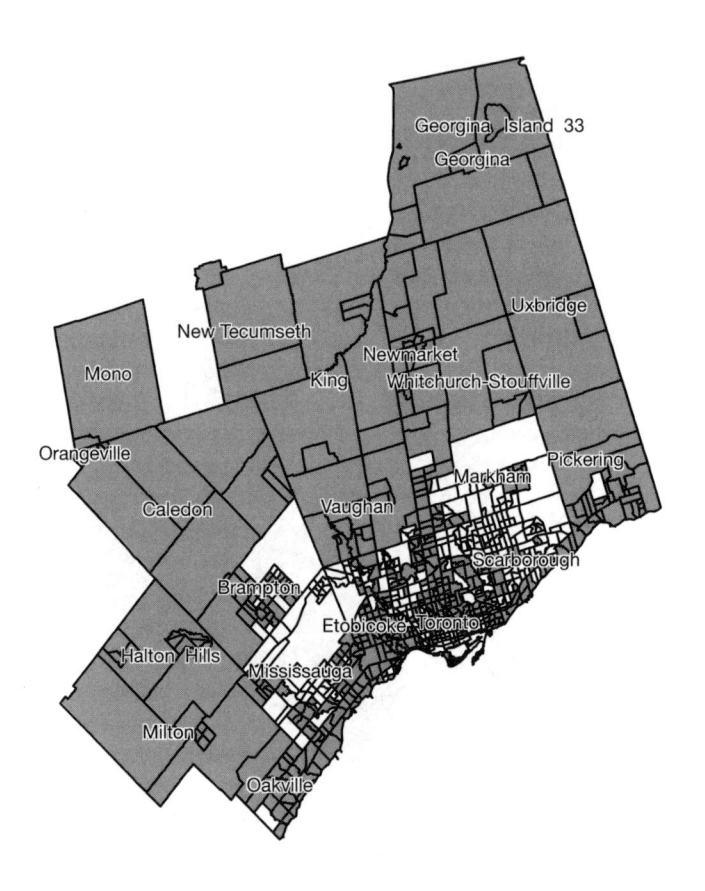

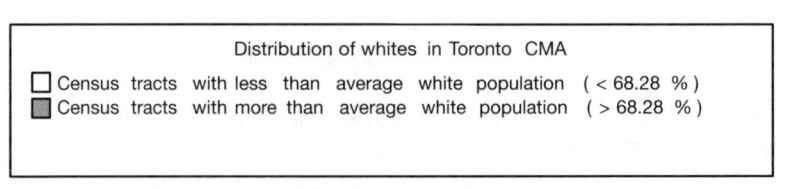

Distribution of whites in Toronto CMA
☐ Census tracts with less than average white population (< 68.28 %)
■ Census tracts with more than average white population (> 68.28 %)

Map 2.2 Distribution of white population across Toronto CMA's census tracts, 1996

means that a majority of the Toronto CMA's Southeast Asian population would have to change places with the white population in order to become desegregated residentially. These findings, however, fall short of telling us anything about the *socioeconomic quality* of the neighbourhoods where the two groups reside and are concentrated. For that answer I examined the socioeconomic characteristics of their neighbourhoods.

Differences in socioeconomic quality of neighbourhoods where whites and Southeast Asians reside in Toronto CMA

This section has implications for whether there are residential, social and economic quality of life *consequences* associated with maintaining group segregation instead of spatial assimilation. The difference in neighbourhood quality of life will be measured here in terms of socioeconomic characteristics of the neighbourhoods where the Southeast Asians and whites live.

The key question is how the two groups are distributed over the different socioeconomic classes in Table 2.1. Table 2.2 illustrates the percentages of white and Southeast Asian groups to the total population, including other groups, for each socioeconomic class. Taking the population of the upper class, for example, the percentage of whites is 88.04 per cent versus 0.10 per cent for Southeast Asians. These figures represent the percentage of each group in the upper-class neighbourhoods with the highest socioeconomic characteristics in comparison with the total population in Toronto CMA. At the other end of the spectrum, the lower-class neighbourhoods have 34.08 per cent whites and 5.10 per cent Southeast Asians.

Table 2.2 also presents the share of whites and Southeast Asians in each socioeconomic class relative to each group's distribution in the Toronto CMA as a whole – that is, it indicates how each white and Southeast Asian group's population is divided over the socioeconomic classes. For example, 3.16 per cent of whites in Toronto CMA reside in upper-class neighbourhoods with the highest socioeconomic index scores, versus only 0.23 per cent of the Toronto CMA's Southeast Asian population. The rest of the whites and Southeast Asians are distributed over the other three lower socioeconomic classes. The lower-class neighbourhoods account for only 1.86 per cent of whites in Toronto CMA, but 17.53 per cent of the Southeast Asian population. As stated previously, where a group lives significantly affects that group's quality of life.

The residential pattern revealed by Table 2.2 demonstrates that, compared with Southeast Asians, a higher percentage of whites in Toronto CMA is concentrated in neighbourhoods with the highest socioeconomic characteristics. More than 3 per cent of all whites in Toronto CMA is part of the super-elite living in upper-class neighbourhoods. They live in neighbourhoods where 65 per cent of the population (fifteen years old and over) have university degrees and the median income of residents is over $100,000. Moreover, more than a third of the residents hold managerial and professional positions; the average value of their homes is more than $500,000 and 72 per cent of this group own

Table 2.1 Socioeconomic indicators for Toronto CMA by quality of neighbourhoods, 1996

Neighbourhoods	Percentage university	Median income	Percentage manager	Average dwelling value	Average dwelling rent	Percentage ownership	Incidence of low income	Unemployment rate
Upper class	65.06	$115,508	34.91	$522,418	$1,194	71.77	5.29	4.73
Upper-middle class	36.06	$66,685	16.55	$268,671	$934	74.36	10.55	6.58
Lower-middle class	22.85	$43,313	9.00	$191,873	$735	52.42	24.31	11.23
Lower class	18.87	$25,157	6.25	$151,774	$591	19.76	51.23	20.30
Toronto CMA	**30.37**	**$56,155**	**13.35**	**$238,426**	**$838**	**62.04**	**18.21**	**9.17**

Source: Calculated by the author from Statistics Canada, 1998.

Table 2.2 Percentage and share of Southeast Asians and whites in Toronto CMA by socioeconomic quality of neighbourhoods, 1996

	Percentage			
Neighbourhoods	Southeast Asians	Whites	Other groups	Total
Upper class	0.10	88.04	11.86	100
Upper-middle class	0.44	77.96	21.60	100
Lower-middle class	1.49	63.72	34.79	100
Lower class	5.10	34.08	60.82	100
	Share			
Neighbourhoods	Southeast Asians	Whites		
Upper class	0.23 %	3.16 %		
Upper-middle class	18.14 %	51.36 %		
Lower-middle class	64.10 %	43.62 %		
Lower class	17.53 %	1.86 %		
Total	100 %	100 %		

Source: Calculated by the author from Statistics Canada, 1998

Segregation and neighbourhood inequality 41

their homes. What is more, these are neighbourhoods where renters are able to pay more than $1,000 per month, which explains why the incidence of low income is only 5 per cent and the unemployment rate 4.7 per cent. Furthermore, more than half of all whites in Toronto CMA live in the upper- and upper-middle-class neighbourhoods – both above average neighbourhoods for the Toronto CMA.

On the other hand, less than 1 per cent of all Southeast Asians in Toronto CMA live in the super-elite upper-class neighbourhoods. Unlike whites, the majority (81.6 per cent) of all Southeast Asians live in the lower- and lower-middle-class areas – that is, below average neighbourhoods in the Toronto CMA. Only 45 per cent of whites live in such neighbourhoods. What is more, 17 per cent of all Southeast Asians live in the poorest neighbourhoods in the Toronto CMA. These are neighbourhoods where only 19 per cent of the population (fifteen years old and over) has university degrees. The median income is only $25,000 and only 6 per cent of the population is in professional and managerial positions. Most of the residents (81 per cent) do not own their homes. Those who do own their homes reside in neighbourhoods where the

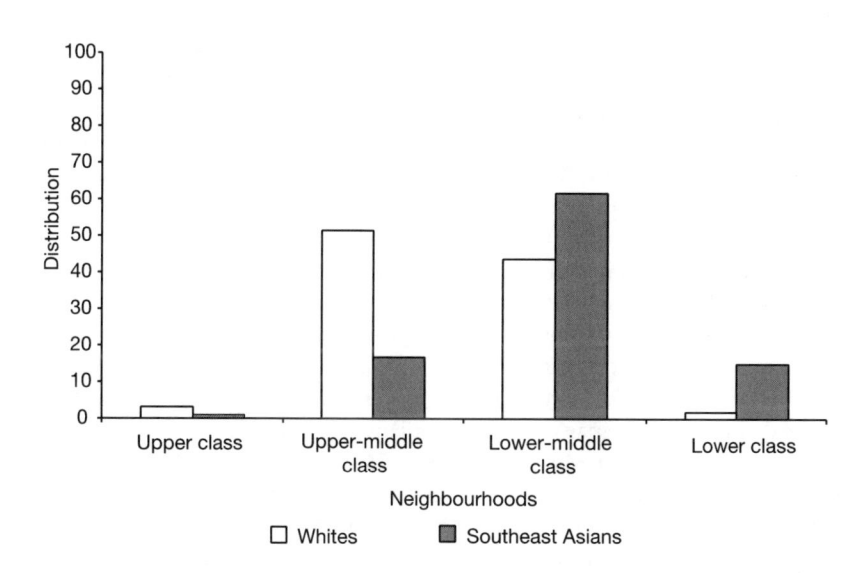

Figure 2.1 Distribution of Southeast Asians and whites in Toronto CMA by socioeconomic quality of neighbourhoods, 1996

average value is $151,000. Most of the residents (51 per cent) have low incomes and the unemployment rate is 20 per cent.

The overall picture of Southeast Asians and whites' representation in each socioeconomic class is presented in Figure 2.1. The graph shows the uneven residential distribution of the two groups across the city's socioeconomic classes

– that is, it indicates how each white and Southeast Asian group's population is divided over the socioeconomic classes. The figure clearly shows *inequality* in the distribution of Southeast Asians and whites residing in high-quality and poor-quality neighbourhoods.

Conclusions

This paper has examined the extent to which Southeast Asians are residentially segregated from the white Canadian population in the Toronto Census Metropolitan Area. It also assesses the social and economic consequences of that segregation by examining the extent of inequality between Southeast Asians and whites as indicated by the characteristics of the neighbourhoods where the two groups reside.

The analysis of 1996 census data from Statistics Canada revealed that Southeast Asians are highly segregated from white Canadians, with an index of dissimilarity of 57 per cent. An index above 50 per cent means that a majority of Southeast Asians would have to change places of residence with whites in order to become desegregated.

One of the consequences of residential segregation is that the characteristics of the neighbourhoods where Southeast Asians reside are grossly unequal to those in the neighbourhoods occupied by whites. These findings have implications for Canada's multiculturalism policy, which promotes cultural retention and enhanced development of communities while also promoting social equality – that is, equal access and participation for all Canadians in the economic, social, cultural and political life of the nation. It appears from these results that Southeast Asians, a visible minority group in a predominantly white society, have not achieved the equal access and social equality goal of the multiculturalism policy.

Equality will not come without persistent struggle. This will involve the formation of political coalitions with other groups and an increased consciousness of the role that racial discrimination plays in limiting Southeast Asians' social and spatial mobility. It will also be a continuous challenge to the well-established myth among many white Canadians that there is no discrimination in Canada (Billingsley and Muszynski 1985). Lessons may be learned from the United States, where people of colour have been struggling for racial equality over a longer period of time. They have formed effective coalitions and have challenged racial discrimination more vigorously, which has resulted in stronger penalties for offences. The result has been a greater measure of success in achieving racial equality in the United States.

Despite the difficulty in achieving racial equality, Toronto is a multicultural and multiracial metropolis. It is continuing to celebrate its diversity. Thus Southeast Asian immigrants should not give up on their dream of a better life. The evidence presented here suggests that a better life can only be achieved through struggle.

Notes

1 In Canada, census tracts are small urban neighbourhood-like communities within census metropolitan areas and census agglomerations that had urban core populations of 50,000 or more in the previous census. The population of census tracts ranges from 2,500 to 8,000, averaging 4,000 in size (except for census tracts in business districts). When census tracts are first delineated, Statistics Canada makes every effort to ensure that the social/economic status of their populations is as homogeneous as possible (Statistics Canada 1997: 195–7).
2 This variable represents the population with a university degree as a percentage of total population fifteen years old and over by highest level of schooling – that is, adults with at least some level of schooling rather than the population as a whole.
3 Categories chosen to be included as managerial and professional status positions are senior management occupations, professional occupations in business and finance, professional occupations in natural and applied sciences, professional occupations in health, professional occupations in art and culture, judges, lawyers, psychologists, social workers, ministers of religion, and policy and programme officers. The variable is calculated as a percentage of the experienced labour force. The experienced labour force refers to people fifteen years old and over, excluding institutional residents, who were employed or unemployed during the week prior to census day and who had last worked for pay or in self-employed activities in either 1995 or 1996. The experienced labour force can be derived by excluding from the total labour force those unemployed people fifteen years old and over who have never worked or who had last worked prior to 1 January 1995 (Statistics Canada 1998).
4 The incidence of low income is the percentage of families or unattached individuals in a given tract below the low income cut-offs. Low income cut-offs are established based on national family expenditure data from 1969, 1978, 1986 and 1992. These data indicate that Canadian families spent, on average, 42 per cent of their income on basic necessities in 1969, 38.5 per cent in 1978, 36.2 per cent in 1986 and 34.7 per cent in 1992. The 1996 low income cut-offs are updated by changes in the consumer price index and differentiated by family size and degree of urbanization (Statistics Canada 1998).

References

Adams, John (1984) 'The meaning of housing in America', *Annals of the Association of American Geographers* 74: 515–26.

Agnew, John (1981) 'Homeownership and identity in capitalist societies', in James Duncan (ed.) *Housing and Identity: Cross-cultural perspectives*, London: Croom Helm.

Alba, Richard and Logan, John (1991) 'Variations on two themes: racial and ethnic patterns in the attainment of suburban residence', *Demography* 28: 431–54.

—— (1993) 'Minority proximity to whites in suburbs: an individual-level analysis of segregation', *American Journal of Sociology* 98(6): 1388–427.

Backer, John C. (1993) *Keeping to the Marketplace: The evolution of Canadian housing policy*, Montreal: McGill-Queen's University Press.

Balakrishnan, T.R. (1976) 'Ethnic residential segregation in the metropolitan areas of Canada', *Canadian Journal of Sociology* 1: 481–98.

Balakrishnan, T.R. and Selvanathan, E. (1990) 'Ethnic segregation in metropolitan Canada', in Shiva S. Hilli, Frank Trovato and Leo Dreidger (eds), *Ethnic Demography*, Ottawa: Carlton University Press.

Balakrishnan, T.R. and Wu, Zheng (1992) 'Homeownership patterns and ethnicity in selected Canadian cities', *Canadian Journal of Sociology* 17: 389–403.

Billingsley, Brenda and Muszynski, Leon (1985) *No Discrimination Here? Toronto employers and the multi-racial workforce*, Toronto: Social Planning Council of Metropolitan Toronto and the Urban Alliance on Race Relations.

Blau, Peter M. (1977) *Heterogeneity and Inequality: A primitive theory of social structure*, New York: Free Press.

Blum, Terry and Kingston, Paul William (1984) 'Homeownership and social attachment', *Sociological Perspectives* 27(2): 159–80.

Briggs, Xavier, Darden, Joe T. and Aidala, Angela (1999) 'In the wake of desegregation: early impacts of scattered site public housing on neighborhoods in Yonkers, New York', *Journal of the American Planning Association* 65(1): 27–49.

Darden, Joe T. (1989) 'Blacks and other racial minorities: the significance of color in inequality', *Urban Geography* 10: 562–77.

Darden, Joe T. and Haney, Jane B. (1978) 'Measuring adaptation: migration status and residential segregation among Anglos, blacks and Chicanos', *East Lakes Geographer* 13(1): 20–33.

Darden, Joe T. and Kamel, Sameh (1999) 'Spatial and socioeconomic analysis of Arab/West Asians and Jews in the Toronto CMA', *The Arab World Geographer* 2(2): 149–73.

Darden, Joe T. and Tabachneck, Arthur (1980) 'Algorithm 8: graphic and mathematical descriptions of inequality, dissimilarity, segregation, or concentration', *Environment and Planning A* 12: 227–34.

Darroch, Gordon A. and Marston, Wilfred G. (1971) 'The social class basis of ethnic residential segregation: the Canadian case', *American Journal of Sociology* 77: 491–510.

Department of Manpower and Immigration (1974) *Canadian Immigration and Population Study: Immigration program*, vol. 2, Ottawa: Information Canada.

Duncan, Otis and Duncan, Beverly (1955) 'Methodological analysis of segregation indexes', *American Sociological Review* 20: 210–17.

Engels, Friedrich (1936) *The Housing Question*, London: Martin Lawrence.

Fong, Eric and Gulia, Milena (1996) 'The attainment of neighborhood qualities among British, Chinese, and black immigrants in Toronto and Vancouver', *Research in Community Sociology* 6: 123–45.

Hannley, Lynn (1993) 'Substandard housing', in John Miron (ed.), *House, Home, and Community: Progress in housing Canadians, 1945–1986*, Montreal: McGill-Queen's University Press.

Henry, Frances (1994) *The Caribbean Diaspora in Toronto: Learning to live with racism*, Toronto: University of Toronto Press.

Hernandez, Marcela (1994) 'Evaluation of long term implications of social policies: the 1967 immigration act as a case study', unpublished M.A. thesis, University of Windsor, Windsor, Ontario, Canada.

House of Commons Debates (1971) Canada: House of Commons Debates.

Massey, Douglas (1985) 'Ethnic residential segregation: a theoretical synthesis and empirical review', *Sociology and Social Research* 69: 315–50.

Massey, Douglas and Denton, Nancy (1993) *American Apartheid: Segregation and the making of the underclass*, Cambridge: Howard University Press.

Minister Supply and Services Canada (1996) *Profiles Vietnam: Immigrants from Vietnam in Canada* (Immigration Research Series), Ottawa: Citizenship and Immigration Canada.

Perin, Constance (1977) *Everything in Its Place*, Princeton: Princeton University Press.

Pratt, Geraldine (1987) 'Class, home and politics', *Canadian Review of Sociology and Anthropology* 24(1): 39–57.

Rakoff, Robert M. (1977) 'Ideology in everyday life: the meaning of the house', *Politics and Society* 7: 85–104.

Ray, Brian and Moore, Eric (1991) 'Access to homeownership among immigrant groups in Canada', *The Canadian Review of Sociology and Anthropology* 28(1): 1–27.

Rose, Eliot *et al.* (1969) *Colour and Citizenship*, London: Oxford University Press.

Rosenbaum, James (1995) 'Changing the geography of opportunity by expanding residential choice: lessons from the Gautreaux program', *Housing Policy Debate* 6(1): 231–70.

Saunders, Peter (1978) 'Domestic property and social class', *International Journal of Urban and Regional Research* 2: 233–51.

Skaburskis, Andrejs (1996) 'Race and tenure in Toronto', *Urban Studies* 33: 223–52.

Smith, Susan (1989) *The Politics of Race and Residence*, Cambridge: Policy Press.

Statistics Canada (1997) *1996 Census Dictionary*, Ottawa: Industry Canada, 1996 Census of Canada, Catalogue no. 92–351–XPE.

—— (1998) *1996 Census of Canada [CD-ROM] – The profile series*, Catalogue no. 95–F0253XCB96000, Ottawa: Ministry of Industry.

Statutes of Canada (1988) *Canadian Multiculturalism Act*, C31.

Sullivan, Oriel (1989) 'Housing tenure as a consumption-sector divide', *International Journal of Urban and Regional Research* 13: 183–200.

Verberg, Norine (2000) 'Homeownership and politics: testing the political incorporation thesis', *Canadian Journal of Sociology* 25(2): 169–95.

West, Cornell (2000) 'Prophetic alternatives: a conversation with Cornell West', in Marable Manning (ed.), *Dispatches from the Ebony Tower: Intellectuals confront the African American experience*, New York: Columbia University Press.

3 Places of worship in multicultural settings in Toronto

Chandrakant P. Shah

Religion plays an important part in people's lives in all societies. A few statistics about the state of religion in the USA today are illuminating. Regarding individual beliefs about 'God', 95 per cent of Americans believe in God, more than 50 per cent pray daily and more than 40 per cent attend church weekly (Puchalski and Larson 1998). Interestingly, almost three-quarters of Americans say their approach to life is grounded in their religious faith. This includes the times when people are ill. Almost three-quarters of patients felt that their physician should address spiritual issues as part of their medical care. At the same time, about 40 per cent of patients wanted their physician to discuss their religious faith with them and almost 50 per cent of patients wanted their physician to pray with them.

In recent years, the literature has increasingly shown that religious individuals have better health in terms of both the length and the quality of their lives. In the *Archives of Family Medicine*, a ten-year review stated that 81 per cent of the studies found religion to have a positive relationship with physical health (Matthews *et al.* 1998). Also, 4 per cent of the studies found religion to be harmful while 15 per cent of the studies found it neutral to the patients' health.

So it is from this perspective that the present study focuses on the impact of religious practices not only on mortality rates of the population (Part I) but also on the spatial distribution of places of worship (Part II) for the religious ethnic minority groups in Toronto, Canada.

In the final analysis, the report implies that religious beliefs can reduce what is viewed as stressful – including inequality and discrimination in a multicultural society – and improve 'coping styles', reducing the risk of decline in physical health. It follows that the existence of a place of worship for ethnic minorities in the vicinity of their dwellings would be beneficial to this end, alleviating their predicaments, whereas the lack of such a facility, or efforts to prevent the construction of religious facilities, increases the inequality experienced by ethnic minorities. Thus minority faiths and religious behaviour affects the emergence and spatial distribution of ethnic minority neighbourhoods.

Part I: Impact of religious practices on mortality rates

Levin identified eight hypothetical pathways by which religious involvement could influence health (Levin 1996). Table 3.1 shows a number of possible mechanisms. Levin makes the point that items in the list are included for completeness and acknowledges that the bottom two pathways (namely, superempirical effects and supernatural effects) are beyond the ability of science to detect. The fourth pathway (psychodynamics of ritual), listed explicitly, identifies religious involvement and fellowship as contributing to social support and stress-reduction. It is possible that other pathways would also influence health in this way. For example, improved health behaviour would improve coping, while better coping strategies in other areas of life might lead to less adverse health behaviour. It is also possible that religious beliefs and worldviews would reduce those situations viewed as stressful and improve coping styles. The psychodynamics of ritual would also probably have stress-reducing or mediating effects.

Table 3.1 Hypothesized pathways by which dimensions of religious involvement influence health

Religious dimensions	Pathways	Mediating factors	Salutogenic mechanisms
Religious commitment	Health-related behaviour and lifestyles	Avoidance of smoking, drinking, drug use, poor diet, unprotected sex etc.	Lower disease risk and enhanced well-being
Religio-ethnic identity	Heredity	Phenotype	Hereditary transmission
Religious involvement and fellowship	Social support	Social relationships, supportive networks, and friends and family	Stress-buffering, coping and adaptation
Religious worship and prayer	Psychodynamics of ritual	Relaxation, hope, forgiveness, catharsis, empowerment, love, contentment and positive emotions	Psychoneuro-immunology, psychoneuro-endocrinology and psychophysiology
Religious and theological beliefs and worldviews	Psychodynamics of belief	Salutary health benefits, personality styles, and behavioural patterns	Consonance between religious and health-related cognitions
Religious faith	Psychodynamics of faith	Optimism and positive expectation	Placebo effect
Religious, spiritual and mystical or numinous experience	Superempirical effects	Activation or invocation of healing bioenergy or a lifeforce and experience of altered states of consciousness	Naturalistic subtle energy and non-local effects
Religious obedience (via faith, behaviour, worship or prayer)	Supernatural effects	Divine blessing	Supernatural effects

Source: Levin 1996

Method of the study for Part I

The Medline and PsychInfo databases were used in associating religion with one of the following: cohort analysis, prospective studies, case-control studies or controlled clinical trials. The studies selected were limited to prospective, case-control and randomized designs using the measurements of exposure risks or types of risk estimates (for example, 'relative risks', 'odds ratios', 'hazard ratios', 'relative hazards').

In this context, 'relative risk' means the ratio of incidence rates of an event or outcome in those exposed compared to those not exposed. On the other hand, 'odds ratio' – an estimate of 'relative risk' – gives the ratio of the odds in favour of exposure among cases to the odds in favour of exposure among non-cases. Another measurement, 'hazard ratio' (HR) – or 'relative hazard' (RH) – provides the estimate of risk from proportional hazards model analysis.

References from retrieved articles were also scanned for articles not otherwise identified. The studies selected had mortality or disease-specific outcomes. Therefore, ecological and cross-sectional studies and studies using other types of analysis (for example, linear regression) were excluded.

Results of the literature review on religiosity and health

An overview of several cohort and case-control studies and two randomized studies is presented below. Table 3.2 is a summary table of the studies and their results.

A   Cohort studies

The Alameda County Study, carried out in California, USA, has been a prospective, community-based study of 6,928 persons since 1965. With the follow-up studies of seventeen years in duration, lack of membership in a church group was associated with all-cause morality in all age groups except for those aged 50–59 (Seeman et al. 1987). For those aged 38–49, 60–69 and 70+, the relative hazards were 1.82 (95% CI 1.27–2.59), 1.40 (95% CI 1.08–1.80) and 1.32 (95% CI 1.08–1.62) respectively. Analysis was controlled for age, sex, race, baseline health status, perceived health, depression and health practices. After twenty-eight years of follow-up studies, frequent religious attendance was associated with lower mortality rates compared with infrequent attendees (relative hazard (RH)=0.64; 95% CI 0.53–0.77) (Strawbridge et al. 1997). Mortality reductions were much greater for females than males. Additional analyses with incremental addition of health conditions, social connections and health practices led to slight increases in the RH with those calculated for males becoming non-significant. Over the course of the study, those that had greater attendance were more likely to increase social connections and to improve health behaviour.

A twelve-year follow-up study of a population cohort from Evans County in Georgia, USA, found that spending spare time on church activities was protective for mortality in white males (hazard ratio (HR)=1.3; 95% CI 1.0–1.9)

Table 3.2 Religion and health outcomes

Source	Design	Measure of religion	Outcome type	Findings
Cohort studies				
Seeman et al. (1987)	Prospective cohort; 17-year follow-up	Church group member	Mortality	38–49 yrs: HR=1.82 (1.27–2.59) 60–69 yrs: HR=1.40 (1.08–1.80) 70+ yrs: HR=1.30 (1.08–1.62)
Strawbridge et al. (1997)	Prospective cohort; 28-year follow-up	Religious attendance (more vs less)	Mortality	RH=0.64 (0.53–0.77)
Schoenbach et al. (1986)	Prospective cohort; 12-year follow-up	Church activities	Mortality	HR=1.3 (1.0–1.9; white males only)
House, Robbins and Metzner (1982)	Prospective cohort; 12-year follow-up	Church attendance	Mortality	RR=1.86 (women only)
Kark et al. (1996)	Historical cohort; 15 years	Secular vs religious kibbutzims	Mortality	RR=1.98 (1.50–2.62)
Goldbourt, Yaari and Medalie (1993)	Prospective cohort; 23 years	Index based on religious or secular education, self-definition and synagogue-going	Coronary heart disease mortality	RR=0.79
Oman and Reed (1998)	Prospective cohort; 4.9 years	Religious attendance (more vs less)	Mortality	RH=0.76 (0.62–0.94)
Colantonio, Kasl and Ostfeld (1992)	Prospective cohort; 6 years	Religious attendance (high vs low)	Stroke incidence	RR=0.86 (0.79–0.94)
Oxman, Freeman and Manheimer (1995)	Prospective cohort; post-open heart surgery; 6 months	Strength and comfort from religion (none vs high)	Mortality	RR=3.25 (1.09–9.72)
Koenig, George and Peterson (1998)	Prospective cohort; depressed patients; 48 weeks	Intrinsic religiosity[1] (low vs high)	Remission from depression	HR=1.7 (1.05–2.75)
Hummer et al. (1999)	Prospective cohort, US sample; 8 years	Religious attendance	Mortality	RR1.87
Koenig et al. (1998)	Prospective cohort; hospitalized veterans	Religious affiliation and religious coping	Mortality	RR=1.00 (0.99–1.01)
Bryant and Rakowski (1992)	Prospective cohort; elderly African-Americans	Religious attendance	Mortality	RR=1.77

Table 3.2 continued

Source	Design	Measure of religion	Outcome type	Findings
Case-control studies				
Friedlander, Kark and Stein (1986)	Case-control	Jewish orthodoxy (secular vs orthodox)	First myocardial infarction	Men: OR=4.2 (1.6–6.6); women: OR=7.3 (2.3–23.0)
Zuckerman, Kasl and Ostfeld (1984)	Nested case-control	Index of religiousness[2] (less vs more)	Mortality	OR=2.02 (CI not provided)
Kune, Kune and Watson (1993)	Case-control	Self-reported 'religiousness' (high vs low)	Colorectal cancer	OR=0.7 (0.6–0.9)
Randomized studies				
Byrd (1988)	RCT; CCU patients (patients and assessor blinded)	Intercessory prayer (those prayed for vs those not)	Complications	RR=0.56[3] (0.37–0.84)

Notes:
1 Scale consisting of ten statements about religious belief or experience.
2 Attendance, self-report religiousness and religion as source of strength.
3 Calculated from table of data provided in reference; 'intermediate' and 'bad' combined for poor outcome; placing 'intermediate' with 'good' generates statistically significant RR=0.64.

(Schoenbach et al. 1986).[1] By contrast, after twelve years of follow-up in the Tecumseh Community Health Study carried out in Michigan, USA, less frequent church attendance was associated with increased mortality in women (relative risk (RR)=1.86) (House, Robbins and Metzner 1982).[2]

A fifteen-year historical cohort study compared the mortality experience of those living in 'secular – i.e., not overtly or specifically religious; indifference to or exclusion of religious considerations – kibbutzim' versus 'religious kibbutzim' (Kark et al. 1996).[3] Overall, those living in secular kibbutzim had significantly increased age-adjusted mortality (RR=1.98; 95% CI 1.50–2.62). Another study from Israel tracked more than 10,000 male civil servants over a twenty-three-year period. A 20 per cent reduction in cardiac deaths was observed in those who considered themselves 'most orthodox' compared with other groups ('orthodox', 'traditional', 'secular', 'non-believers') (Goldbourt, Yaari and Medalie 1993).[4]

In a cohort study (follow-up studies averaged 4.9 years) of community-dwelling residents aged 55 and over, living in Marin County, California, USA, religious attendance was found to reduce the risk of mortality after controlling for age, sex, physical functioning and social support (RH=0.76; 95% CI 0.62–0.94) (Oman and Reed 1998). Religious attendance was slightly more protective for respondents with high levels of other social support.

In a follow-up study of patients undergoing open heart surgery (Oxman, Freeman and Manheimer 1995), patients stating that they received no strength and comfort from religion were over three times more likely to die in the six-month post-operative period after controlling for biomedical indices (RR=3.25; 95% CI 1.09–9.72; p=0.04). Lack of social participation (group activities) was also a significant factor.[5]

Religious attendance and subsequent mortality was investigated in a random sample of the population in the USA (1987 National Health Interview Survey, see Hummer et al. 1999). Mortality over an eight-year period was found to be increased in those who never attended church, temple or other religious services compared with those who attend more than once a week, (RR=1.87). The effect of religious attendance appeared to be mediated by healthier behaviours and social ties. In comparison, in a longitudinal study of aging, the mortality of 473 elderly African-Americans was found to be associated with church attendance (RR=1.77) (Bryant and Rakowski 1992).[6] Comparing men and women, men who did not attend church had the highest risk of death (RR=2.72) compared with women who did attend church in multivariate analysis.

B   Case-control studies

A case-control study from Israel compared Jewish patients with their first myocardial infarction, with Jewish controls without a history of coronary artery disease. All subjects were asked to self-report their level of religious orthodoxy ('orthodox', 'traditional' or 'secular'). After controlling for age, ethnicity, education, smoking, physical activity and BMI, secular subjects had a significantly higher risk of MI compared to orthodox subjects (men: OR=4.2, 95% CI 2.6–6.6; women: OR=7.3, 95% CI 2.3–23.0) (Friedlander, Kark and Stein 1986).[7] Likewise, a case-control study from the Melbourne Colorectal Cancer Study found that self-reported 'religiousness' was associated with a significantly reduced risk of colorectal cancer (OR=0.7, CI 0.6–0.9, p=0.002) (Kune, Kune and Watson 1993).[8]

A study by Zuckerman, Kasl and Ostfeld (1984) performed a secondary analysis on a two-year cohort study comparing the health impact of mandatory relocation of urban poor. Lower levels of religiousness were associated with a higher risk of death (OR=2.02). Neither confidence intervals nor p-values were provided. The index of religiousness was constructed based on attendance at religious services, self-reported religiousness and religion as a source of strength.

C   Randomized design

In a blinded randomized study by Byrd (1988)[9] of the impact of intercessory prayer (IP) on patients admitted to a coronary care unit, patients who received IP were much less likely to experience complications during their stay (RR=0.56 (calculated), 95% CI 0.37–0.84) (Harris et al. 1999).[10] A recent meta-analysis of twenty-nine studies of religious involvement and mortality found that those less frequently involved in religious activity were at increased risk of death (OR=1.29, 95% CI 1.20–1.39) (McCullough et al. 2000). The studies used in the analysis had substantial variation in the measures of religious involvement, populations studied and mixtures of disease-specific and all-cause mortality outcomes.

As observed from the fair and robust evidence, given above, religious involvement, particularly attendance at the place of worship, has a beneficial impact on people's health. With increasing migration from Asia and the Middle East to North America of people with religious backgrounds that differ from the Judeo-Christian faith traditional to North America, the question that needs to be asked is whether they are well served with regards to accessible places of worship in the large metropolitan areas where they usually settle. Thus the second part of this paper presents the geographic mapping of people with minority faiths and their places of worship, as well as documenting some of the struggles these groups have experienced in erecting their places of worship in the Canadian city of Toronto.

Part II: Spatial relationship of place of worship and population distribution of non-Judeo-Christian communities

The Toronto region has historically been a strong immigrant reception area. It has a population of 2.4 million people of which 46 per cent are of non-European origin (officially referred to as 'visible minority'), representing some 110 countries. The countries of origin for immigrants to Toronto have significantly changed over time. Over the past thirty-five years, the number of people coming from European countries has declined steadily, while the number coming from Asian countries has increased. Before 1980, for example, 60 per cent of the immigrants originated in Europe. Since 1980, Asians have become predominant, accounting for more than half of the new arrivals to the Toronto area. Since 1991, the largest number of new immigrants originated in Sri Lanka, followed by the People's Republic of China, the Philippines, Hong Kong and India.

This change in the composition of immigrants suggests a change in the religious needs of the population. If the incorporation of religion or spirituality in one's life is a major factor in health, then the issue of access to place of worship becomes important. In the econd part of this paper, the Christian and Jewish faiths were not taken into consideration as these denominations provide good access to places of worship. However, many places of worship for recent immigrants are either hidden in industrial complexes – as they cannot afford the high prices charged in residential areas – or are far from the city, requiring a great deal of travel.

Method of study for Part II

Using Geographic Information Systems (GIS), the addresses of the places of worship for the Baha'i, Buddhist, Hindu, Islamic, Jain, Sikh and Zoroastrian religions were plotted as points on a map of Toronto and its surrounding regions. Next, using 1991 Canadian census data (the 1996 Census did not inquire about religion), the locations of Buddhist, Hindu, Islamic and Sikh community residents were plotted on the same map with places of worship (the Baha'i, Jain and Zoroastrian residents were too small in number to be plotted). Only the Toronto locations of these minority populations were available; the regions surrounding Toronto were not included. In addition, the number of people in the four faiths was underestimated, since the figures were only available for 1991. Thus a spatial relationship can be seen between the residences of religious members and their existing places of worship in Toronto. The following maps provide a visual means to judge a religious community's access to their places of worship.

The Buddhist community (Map 3.1) is heavily concentrated in the southern part of Metro Toronto and has a good concentration of temples there. However, slight concentrations of Buddhists in the northwest of Toronto and in the east (Scarborough) do not look as well served.

The Islamic community (Map 3.2) looks fairly well served in Toronto. The community is quite spread out and has mosques corresponding to this population pattern. It should be mentioned that Muslims have the greatest number of places of worship compared to all of the minority faiths investigated in this study.

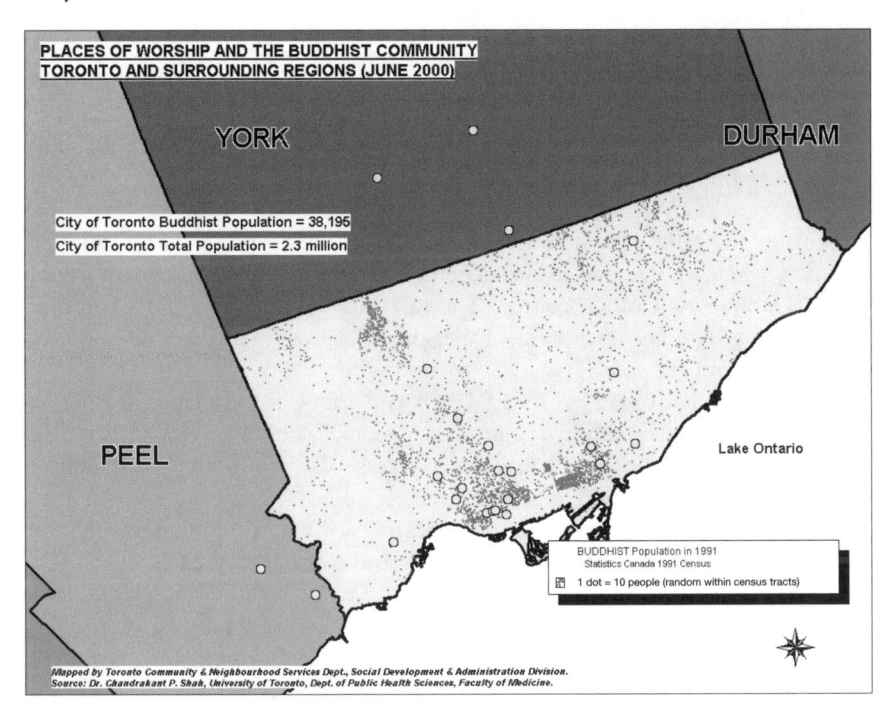

Map 3.1 The Buddhist community in Toronto

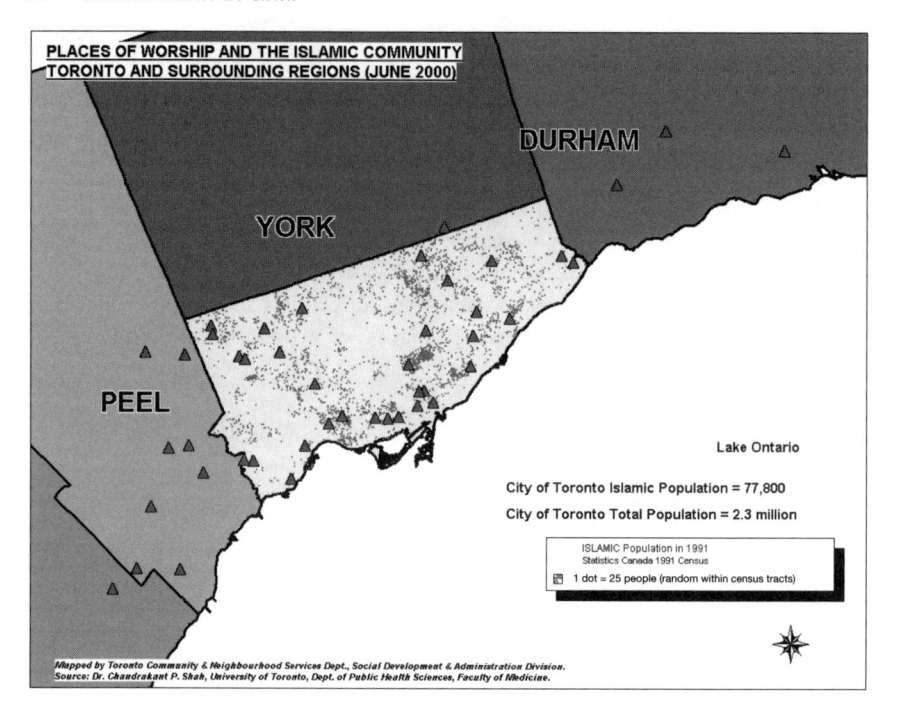

Map 3.2 The Islamic community in Toronto

The Sikh community in east Toronto (Map 3.3) appears to be the least served. Also, the great concentration of Sikhs in the northwest corner of Toronto may demand more than two *gurudwara*s. However, there are three other *gurudwara*s in the neighbouring region of Peel that this community may attend.

Finally, the Hindu community (Map 3.4) has a fairly good concentration of temples where the residents live. There is a greater concentration of people and temples in east Toronto. Though the information for Toronto's surrounding regions is lacking, I personally know that there are a great number of temples.

As Toronto has expanded in the last two decades, buying land for religious purposes for newly established immigrant groups has become very difficult due to exorbitant land prices. Hence many places of worship are being built in outlying areas where land is still reasonably priced. However, even in these areas, minority faith groups meet obstacles from the local resident groups who belong to Judeo-Christian faiths. Two case studies are presented below to illustrate the point.

A case study of a Hindu temple – its construction was blocked by the residents in its neighbourhood – illustrates the spatial segregation and social tension existing in Toronto. Another case study describes the solution found, by a Muslim community, to poor access to their mosque.

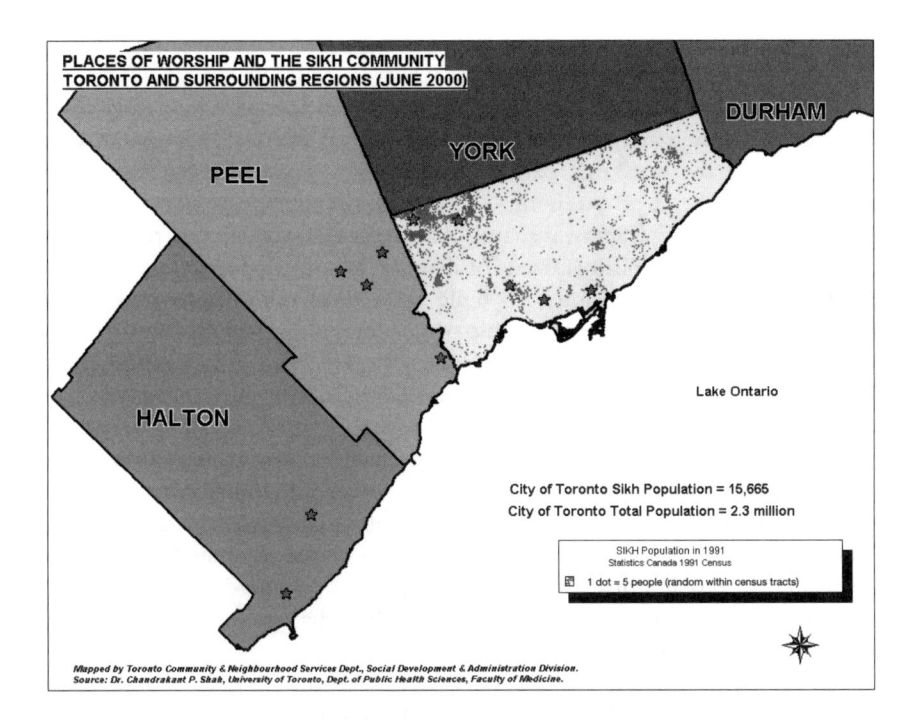

Map 3.3 The Sikh community in Toronto

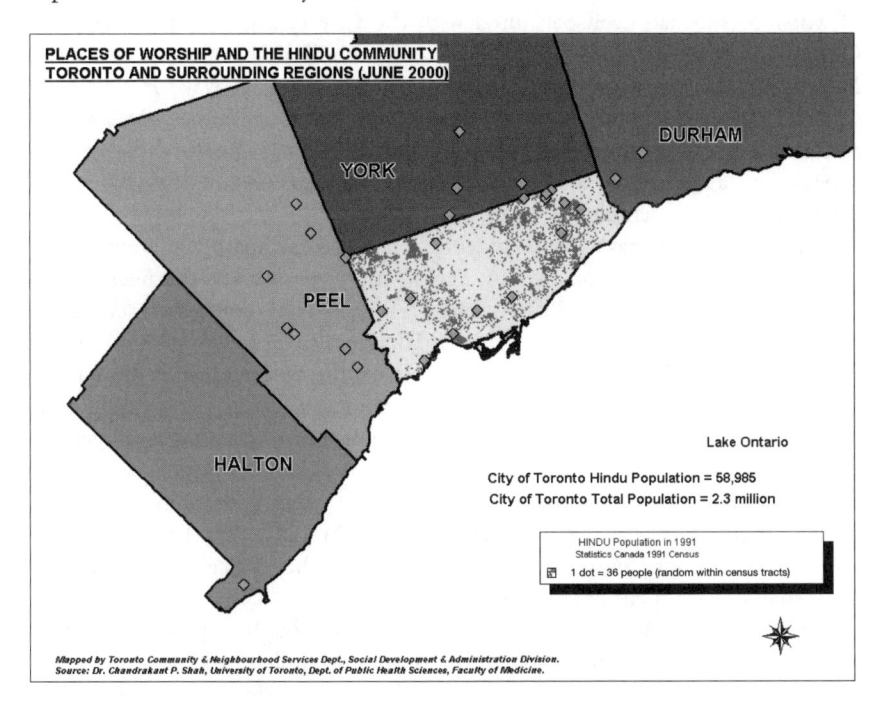

Map 3.4 The Hindu community in Toronto

Case study 1: Hindu temple

In 1976, there were approximately 10,000 Gujaratis (people originating from the state of Gujarat in India) spread all over Metro Toronto, the majority of them being in the northeast part of the city. On weekends, many of these people met to pray at private homes, community centres or in school meeting halls. The community leaders, realizing that a place of worship was required to meet the cultural and spiritual needs of its residents, developed different strategies to raise the necessary funds to build the temple. The community would have preferred their place of worship to be nearer to where they lived in Toronto, but this was not feasible financially. Hence, in April 1985, they bought a ten-acre plot of land away from the City of Toronto, in the town of Markham.

In 1965, in its masterplan, the City of Markham had designated this land for a place of worship. Rich people, mainly of Christian faith, inhabited the neighbourhood around this land. Once the surrounding community realized that a Hindu temple would be erected in their neighbourhood, they started lobbying their municipal politicians against such a plan. In August 1985, the City Planning Department imposed an interim bylaw that prohibited building a place of worship at this site.

The Hindu community was naturally upset with the decision and appealed against the bylaw, asking for an exemption. They argued that they had bought the land in good faith on the understanding that this land was zoned for a place of worship, and had even consulted with the City Planning Department. The Gujarati community mounted their own political campaign. In 1988, Markham's Planning Committee approved rezoning to pave way for the 42,000 square foot Hindu temple and cultural centre. The Committee's decision was reaffirmed by the City Council. However, the local ratepayers' associations appealed to a higher authority, the Ontario Municipal Board (OMB), to halt the building of the temple.

During this upheaval, the mayor, who was sympathetic towards the ratepayers, lost the election. A new mayor, more sensitive to the needs of the Gujarati community, was elected. He personally intervened and settled the dispute before the case went to the OMB. He also promised the ratepayers' associations that there would be no temple on the proposed site. He had the site rezoned so that the Gujarati community could sell it as ten separate estate lots to a developer or to individuals, and promised to help find an alternative site for their temple. With the help of the mayor, the community bought five acres of land in 1993, far from the original site, near a shopping plaza, and finished building their place of worship in 1996. In spite of the happy ending, it took the community almost twenty years to build their place of worship, with more than decade spent entangled in legal battles.

Similarly, in 1996, a group of Islamic followers bought a plot of land for a mosque in the City of Toronto and found opposition from the neighbourhood and municipal counsellors. Fears expressed by the opposing group pertained to the parking issue. However, media reports commented on the religious intolerance. In Toronto, the struggle over building places of worship for the minority faiths still continues. One Islamic group has found an innovative solution, as outlined in Case Study 2.

Case study 2: Peace Village

The Peace Village in Vaughan (a suburb of Toronto) is believed to be the first development in North America where an entire community has been built to cater for the religious needs of Islamic residents. For Muslims, a mosque is the focal point of their lives, and in this community it truly has become that. The Bai'tul Muslim community was unable to obtain any land in the City of Toronto. Thus they built the Bai'tul-Islam Mosque, in 1992, in Vaughan – an area that was at the time surrounded only by cornfields. The trek was lengthy for most worshippers, sometimes taking more than an hour. Some people spent more time on the highway than they did worshipping in the mosque. The community then decided to approach a developer with the idea of constructing semi-detached houses on the twenty-hectare site next to the mosque. The developer agreed and was pleased that the housing units were sold out almost immediately. As of May 2000, 150 families had already moved in.

The residents of Peace Village have only positive things to say about this settlement. Muslims are required to pray five times a day. Being so close to the mosque allows them to pray with fellow residents and follow their religion more closely. This also gives a stronger sense of community. Neighbours who share the same kinds of values and beliefs can spend time with one another. The streets have been named after Muslim leaders and celebrities, which reminds residents not only of the contributions and achievements of the Muslim people but also of their roots and history. The settlement makes social occasions more convenient, especially since extended families tend to live under one roof. The land south of the mosque has already been zoned for a shopping plaza, a recreational park, a community centre and a school. The residents of Peace Village strongly believe that they are not isolating themselves but preserving their values, traditions and religious faith to pass them on to the next generation. They are happy to be a part of Canada's multiculturalism, besides being pleased to have found a solution to their problems concerning religious worship.

In summary, attendance at a place of worship is associated with lower mortality and better health. With changing demographic patterns in major cities in North America, attention needs to be paid to the spatial consequences of these shifts. At present, in the large metropolitan areas of North America, it appears that non-Judeo-Christian faith groups are having difficulty securing places of worship near to where they live. Scant attention has been paid by city planners to accommodating the spiritual and religious needs of the visible minorities.

Acknowledgements

I would like to thank Dr Brent Moloughney and Ms Bonnie Lee for their assistance in the preparation of the manuscript, and Mr Harvey Low from the City of Toronto in preparing the maps for this study.

Notes

1 Associations were not significant for white females or African-Americans of either sex. The analysis was also controlled for age, coronary risk factors, social class and physical activity.
2 The analysis was controlled for age, coronary heart disease, FEV1, hypertension and bronchitis.
3 A reduction in mortality rates was observed for both sexes and across diagnostic categories. Little difference in conventional risk factors was observed between the two groups. The economic status of the kibbutzims was also similar.
4 Analysis was controlled for lifestyle factors. No confidence interval was reported.
5 In contrast, in a six-year prospective study of strokes, univariate analysis found that frequency of religious attendance was statistically associated with strokes (RR=0.86; 95% CI 0.79–0.94). Neither religiousness nor religion as a source of support were associated with stroke incidence. Following control for age, diabetes, hypertension, smoking and physical function, religious attendance failed to retain its significance (Colantonio, Kasl and Ostfeld 1992). Likewise, in a study of the survival of elderly patients initially admitted to a general medicine or neurology service bed, neither religious affiliation nor religious coping were found to be associated with survival. It was postulated that the mortality forces of age, underlying medical condition and severity of health problems overwhelmed the weaker effects of psychosocial variables (Koenig, George and Peterson 1998).
6 There was control in the analysis for age, sex, education, physical health and social network.
7 The control group selected may not have been appropriate since it was participants from a lipid prevalence study. Controls tended to be younger, less likely to be of European origin, better educated, much less likely to smoke, much more active and much less likely to be secular. There was no control for socioeconomic status nor blood pressure or lipid levels.
8 The 715 cases were age- and sex-matched to 727 community controls. Analysis also controlled for the previously determined risk factors in the study (family history of colorectal cancer, dietary risk factors, beer consumption, number of children and age at birth of first child). Religiousness was assessed using a single question with a four-point Likert scale.
9 Statistically significant differences were observed for congestive heart failure, diuretic use, cardiopulmonary arrest, pneumonia, antibiotic use and intubation/ventilation. The assessing clinician and patients were blinded to whether a patient was being prayed for by intercessors. Each intercessor was asked to pray daily for a rapid recovery and for prevention of complications and death.
10 A recently published study (Harris *et al.* 1999) from a different set of investigators also performed a randomized controlled trial of remote intercessory prayer on outcomes of Coronary Care Unit patients. Patients were randomly assigned to a prayer or non-prayer group based on the last digit of their hospital number. Neither patients nor staff were aware of the existence of the study. A blinded investigator performed abstracting of charts. Prayer was for a speedy recovery with no complications. A new scoring system was developed for patient outcomes. Prayer began within 1.2 days after CCU admission and continued for a period of 28 days. There was an 11 per cent (p=0.04) reduction in weighted scores in the prayer group versus

the usual care group. No difference in length of CCU stay or hospital stay was observed. Measuring outcomes as in the study by Byrd, found a 13 per cent improvement in the intervention group, although this was not statistically significant.

References

Bryant, S. and Rakowski, W. (1992) 'Predictors of mortality among elderly African-Americans', *Research on Aging* 14: 50–67.

Byrd, R.C. (1988) 'Positive therapeutic effects of intercessory prayer in a coronary care unit population', *Southern Medical Journal* 81: 826–9.

Colantonio, A., Kasl, S.V. and Ostfeld, A.M. (1992) 'Depressive symptoms and other psychosocial factors as predictors of stroke in the elderly', *American Journal of Epidemiology* 136: 884–94.

Friedlander, Y., Kark, J.D. and Stein, Y. (1986) 'Religious orthodoxy and myocardial infarction in Jerusalem – a case-control study', *International Journal of Cardiology* 10: 33–41.

Goldbourt, U., Yaari, S. and Medalie, J.H. (1993) 'Factors predictive of long-term coronary heart disease mortality among 10,059 male Israeli civil servants and municipal employees', *Cardiology* 82: 100–21.

Harris, W.S. *et al.* (1999) 'A randomized, controlled trial of the effects of remote, intercessory prayer on outcomes in patients admitted to the coronary care unit', *Archives of Internal Medicine* 159: 2273–8.

House, J.S., Robbins, C. and Metzner, H.L. (1982) 'The association of social relationships and activities with mortality: prospective evidence from the Tecumseh community health study', *American Journal of Epidemiology* 116: 123–40.

Hummer, R.A. *et al.* (1999) 'Religious involvement and US adult mortality', *Demography* 36: 273–85.

Kark, J.D., Shemi, G., Friedlander, Y., Martin, O., Manor, O. and Blondheim, S.H. (1996) 'Does religious observance promote health? Mortality in secular vs religious kibbutzim in Israel', *American Journal of Public Health* 86: 341–6.

Koenig, H.G., George, L.K. and Peterson, B.L. (1998) 'Religiosity and remission of depression in medically ill patients', *American Journal of Psychiatry* 155: 536–42.

Koenig, H.G. *et al.* (1998) 'Religion and survival of 1,010 male veterans hospitalized with medical illness', *Journal of Religion and Health* 37: 15–29.

Kune, G.A., Kune, S. and Watson, L.F. (1993) 'Perceived religiousness is protective for colorectal cancer: data from The Melbourne Colorectal Cancer Study', *Journal of the Royal Society of Medicine* 86: 645–7.

Levin, J.S. (1996) 'How religion influences morbidity and health: reflections on natural history, salutogenesis and host resistance', *Social Science and Medicine* 43: 849–64.

McCullough, M.E., Hoyt, W.T., Larson, D.B., Koenig, H.G. and Thoresen, C. (2000) 'Religious involvement and mortality: a meta-analytic review', *Health Psychology: Official Journal of the Division of Health Psychology, American Psychological Association* 19: 211–22.

Matthews, D.A. *et al.* (1998) 'Religious commitment and health status: a review of the research and implications for family medicine', *Archives of Family Medicine* 7: 118–24.

Oman, D. and Reed, D. (1998) 'Religion and mortality among the community-dwelling elderly', *American Journal of Public Health* 88: 1469–75.

Oxman, T.E., Freeman, D.H. and Manheimer, E.D. (1995) 'Lack of social participation or religious strength and comfort as risk factors for death after cardiac surgery in the elderly', *Psychosomatic Medicine* 57: 5–15.

Puchalski, C.M. and Larson, D.B. (1998) 'Developing curricula in spirituality and medicine', *Academic Medicine: Journal of the Association of American Medical Colleges* 73: 970–4.

Schoenbach, V.J., Kaplan, B.H., Fredman, L. and Kleinbaum, D.G. (1986) 'Social ties and mortality in Evans County, Georgia', *American Journal of Epidemiology* 123: 577–91.

Seeman, T.E., Kaplan, G.A., Knudsen, L., Cohen, R. and Guralnik, J. (1987) 'Social network ties and mortality among the elderly in the Alameda County study', *American Journal of Epidemiology* 126: 714–23.

Strawbridge, W.J., Cohen, R.D., Shema, S.J. and Kaplan, G.A. (1997) 'Frequent attendance at religious services and mortality over 28 years', *American Journal of Public Health* 87: 957–61.

Zuckerman, D.M., Kasl, S.K. and Ostfeld, A.M. (1984) 'Psychosocial predictors of mortality among the elderly poor: the role of religion, well-being, and social contacts', *American Journal of Epidemiology* 119: 410–23.

4 The impact of local and regional planning on Arab towns in the 'Little Triangle', Israel

Thabet Abu Rass

Introduction

On 9 November 2000, the Prime Minister of Israel, Ehud Barak, announced the establishment of a state commission of inquiry to investigate the violent clashes in early October 2000 in Israel's Arab sector. Barak's decision came after heavy pressure from the leaders of the Israeli Arab community. The government's announcement stated that the commission would investigate the sequence of events that culminated in the killing of thirteen Arab citizens by the security forces of Israel. The violence was triggered by the Temple Mount visit by the opposition leader, Ariel Sharon, and the subsequent *Al Aqsa Intifada* uprising in the occupied territories of the West Bank and in the Gaza Strip. The Arab population was protesting against civic inequality and the excessive use of force by police in cracking down on the demonstrations.

Following the unprecedented confrontations between Arab citizens and the security forces of Israel, twenty-five experts and researchers submitted a comprehensive report to the prime minister, calling for new governmental policy with regard to the country's Arab population. Prepared in response to the disturbing events, the report covers six different aspects of government policy towards Israeli Arabs. The first of these six position papers, incorporated in a new report criticizing Israeli policy towards the country's Arab population, relates to the sensitive topic of land rights and planning.

Arab–Jewish relations have been marked by ethnic competition over the control of land since and even before the establishment of Israel. This competition must be understood within the historical context of the Arab–Israeli conflict. Land expropriation and the discriminatory policies of the Israeli central government and its institutions left the Arab citizens of Israel politically and economically at a disadvantage.

In the early days of statehood, the Israeli government concentrated on consolidating land and creating Jewish agricultural settlements in all-Arab areas within Israel, including the 'Little Triangle' (Map 4.1). These settlements were inhabited by a very limited number of Jews, but they controlled a vast area in the region. In recent years – especially after the massive Jewish immigration from the former Soviet Union – the region of the Little Triangle has witnessed

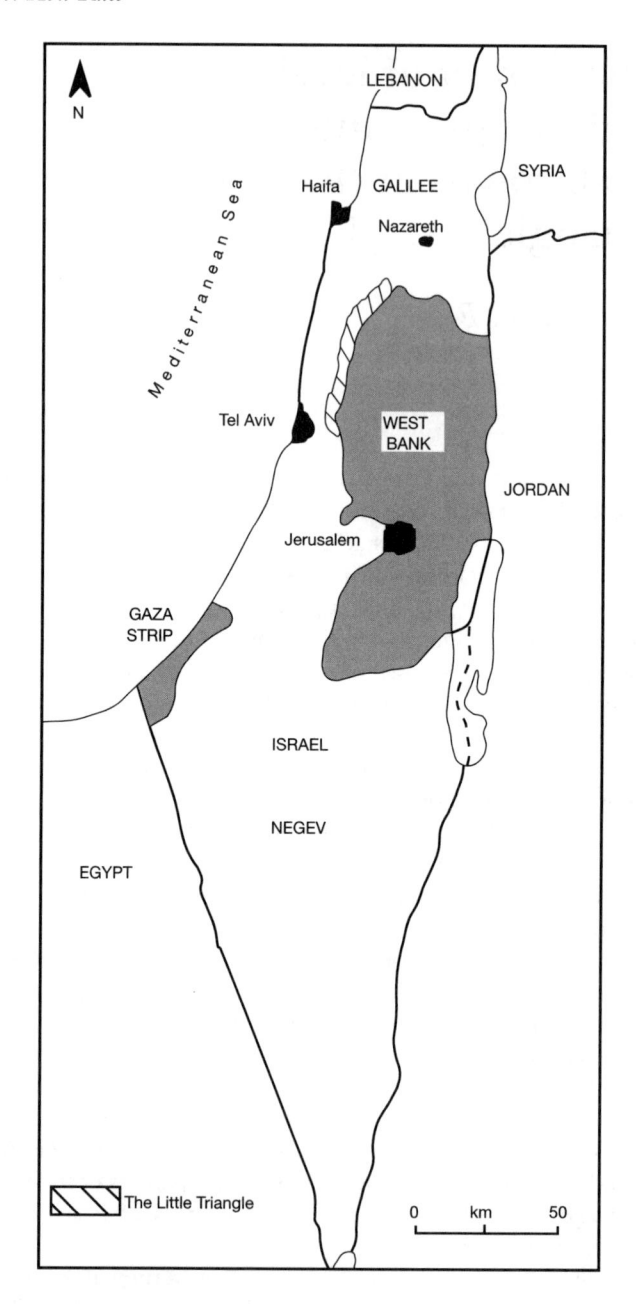

Map 4.1 The Little Triangle region in Israel

an effort to create a spatially fragmented demographic mixture. The creation of several Jewish towns, the enlargement of existing ones and the planned construction of the Trans-Israel Highway will unbalance the demographic order. It will increase the fragmentation of the originally uniform cultural Arab region. According to governmental plans, the area of the Little Triangle will have more Jewish inhabitants than Arabs by the year 2020.

The concentration in state hands of much of the land (93 per cent) in the country also has been an attempt to limit Arab land ownership. Today, the Arab population of Israel owns less than 4 per cent of the total area of Israel. In addition, the state, through different tactics including the allocation of jurisdiction to Arab local government, seeks to control and minimize the amount of land that remains under Arab ownership. The questions at the core of this chapter are:

1 How do land and planning policies and land allocation affect the Arab minority in Israel?
2 Are these policies increasing disparities and uneven development in the Little Triangle?
3 How do the Arab representatives or leaders in the study area perceive the decision-making process, the construction of Jewish settlements in the region and Jewish–Arab relations in the study area?
4 What is the impact of Judaization on the Arab towns and population?

This chapter will show that the Arabs of the Little Triangle are concerned about the reduction in their physical space. They trust neither the decision-making process nor the bureaucracy they are dealing with. They accuse the central government of attempting to sustain the economic and political disparity between Arabs and Jews. This analysis is based on questionnaires and in-depth interviews conducted in 1996.

Theoretical framework

In the literature on planning, the most widespread methodology for evaluating planning programmes has been cost–benefit analysis. In this analysis, the anticipated benefits to be generated by a specific programme are compared with the expected cost (Schofield 1987). The costs and benefits are itemized in terms of monetary value. This methodology is limited because it ignores the social costs and benefits, such as the destruction of a beautiful landscape or important historical buildings, which cannot be always evaluated in financial terms (Cadwallader 1996).

Allen Scott (1980: 37) has suggested three methods of planning strategies used by the state or national planning agencies in urban and local settings. These methods involve the use of fiscal policies, land-regulation policies and development policies. Recently, criticism of mainstream planning theory has led to the emergence of alternative planning theories which focus on the role of public participation in the planning process and emphasize the fact that planning can be viewed as a political act of redistribution (Cadwallader 1996).

Equally important, ethnicity and ethnic groups are closely connected to urban planning and political processes, as multi-ethnic states comprise the overwhelming majority of the world's political units (Soffer 1983). Ethnic conflicts have the potential of threatening internal stability and increasing the instability of state frontiers.

There are four urban planning strategies in ethnically divided societies. The first strategy is a 'neutral' one that is characterized by the de-politicization of territorial issues to avoid political exclusion and inequality of power. In this strategy people are treated as individuals rather than members of an ethnic group. The second urban planning strategy has been labelled 'partisan' by Oren Yiftachel (1995) because it chooses sides. Here the decision-makers are members of the dominant ethnic group. The third strategy is that of 'equality', which tries to minimize the disparities in ethnic states. Thus allocation of resources is largely based on the size and the need of the ethnic group. Finally, there is the 'resolver' urban strategy, which promotes mutual empowerment and a tolerable co-existence in ethnically divided societies (Yiftachel 1995).

The literature that deals with ethnicity and urban development displays various perspectives. The first approach sees ethnicity as a pre-modern cleavage, which will vanish with modernization. Urbanization is perceived as transcending ethnicity (Geertz 1963). The second approach regards ethnicity as an unavoidable component in urban politics. Other scholars confront the issue of ethnicity in a different way. Allen Ross (1982) examines ethnicity as a modern political phenomenon that develops in particular environments and changes with circumstances. Ethnicity is a collective identity, which may appear in different stages of social development and urban growth. Amnon Cohen (1965) suggests that urban ethnic groups are interest groups that are involved in a struggle with other groups for resources in the public arena. Ethnicity functions in the struggle as providing a means of expression that advances solidarity as a moral duty.

The Arabs in Israel

As an outcome of the 1948 Arab–Israeli war, only 156,000 Arabs remained within the boundaries of the newly established state of Israel. The Arabs who were granted Israeli citizenship became, some fifty years later, a sizeable minority of over one million people. Excluding East Jerusalem, this is about 17 per cent of the total population of Israel (Israel Bureau of Statistics 2000a).

The Palestinian Arab minority in Israel has distinctive demographic characteristics, such as a high birth-rate leading to a natural increase in population of 2.9 per cent annually, which is almost twice the rate of the Jewish natural increase. With 45 per cent of the total Arab population in Israel under the age of fifteen, it is a youthful population, like most of the Arab world (Israel Bureau of Statistics 2000b). In addition, Palestinian Arabs have a low rate of emigration. The ratio of Jews to Arabs has remained about the same since the establishment of Israel. The ratio was 1:5 in 1949 and fifty years later it was 1:6. The slight change can be traced to the recent wave of Jewish immigration to Israel from the former Soviet Union. Otherwise, in the last five decades, the natural increase of the Arab population has balanced the total population growth among Jews.

The Arabs, by far the largest ethnic minority in Israel, are divided by three religious affiliations: 76 per cent are Muslims, 14 per cent are Christians and 10 per cent are Druzes (Ghanem 1993). Geographically, the Israeli Arabs are overwhelmingly concentrated in the peripheral areas of Israel: 60 per cent are living in the mountainous Galilee region in northern Israel, 20 per cent in the Little Triangle, 10 per cent in the arid Negev region in southern Israel and the remaining 10 per cent in mixed Israeli cities. Such cities are Acre, Haifa, Jaffa, Lydda and Ramlah (Ghanem 1993).

In the early years of statehood, the Arabs in Israel were perceived as a 'security problem'. They were regarded as a small part of the Arab world and a potential 'fifth column' (Haider 1995). The first prime minister of Israel, David Ben-Gurion, was quoted as saying 'the Arabs have to be judged according to what they could do and not according to what they have actually done' (Lustick 1980: 78). In 1948 the government of Israel imposed a military government on the Arabs that was not officially abolished until 1966. During this period, the movements of the Arabs were restricted and they were excluded from many aspects of mainstream social life in the country. At the same time, a series of land laws was enacted to guarantee full Jewish control over the land gained or 'abandoned' by Palestinians during the war.

The 1967 Six Day War and its outcomes renewed the connection between the Arabs in Israel and those in the West Bank and the Gaza Strip. During the 1970s, the Arabs within Israel proper, now free of the restrictions of the military government, started to form their own organizations and make demands on the Israeli government.

During the 1980s and 1990s, the powerless Arabs started to directly influence the political system in Israel. This period witnessed the emergence of two Arab-nationalist parties, the Progressive Movement for Peace (PMP) and the Arab Democratic Party (ADP). The Islamic Movement in Israel, which was also established in the early 1980s and gained support in the municipal elections, has been participating in parliamentary elections since 1996.

Land and local development policy in Israel

Land in Israel is highly valued by both Arabs and Jews, far beyond its mere economic value. Both national groups perceive it as the basis of national existence. The government land policy in Israel has its roots in ideology and has been used as a tool to achieve the national goal of a Jewish state.

The Israeli local development policy shares some threads with local development and planning policy in Britain. The latter has been justified by securing social justice and sustaining equal standards of living in different regions of the country. In addition, the British encouraged economic effectiveness to enable each region in the country to contribute positively to the national economy (Keeble 1976; McCrone 1969). Similarly, in the Israeli case, the local development and planning policy encouraged social justice, at least among Jews. However, it differs from British local development policy in terms of the highly centralized role played by the state in planning and local development. In the case of Israel, the private sector has little input in local development and planning policy.

One of the earliest Zionist organizations established in Palestine was the Jewish National Fund (JNF), formed in 1901 by the Zionist Congress. The JNF was in charge of purchasing land in Palestine and developing it for Jews. In 1947, the Palestinian Arabs owned over 85 per cent of the total area of Palestine, the Jews 7 per cent and the remaining area was public domain. Following the establishment of Israel in 1948, most of the land that came under the control of the state (over 90 per cent) became publicly owned. The remaining private land was located around cities and Arab towns; most was owned by the Arab citizens of Israel. Public land in Israel cannot be sold, in keeping with the command from the Jewish Holy Book, the Torah: 'The land shall not be sold forever ... ' (Leviticus 25: 23). In 1960, the government of Israel established the Israeli Land Administration (ILA) and took charge of all public land in the country. By 1999, it publicly owned some 93 per cent of the total land in Israel.

Publicly owned land is leased for forty-nine years with a renewal clause. Householders, private farmers, industrialists and collective and cooperative settlements have the right to lease land from the ILA. Their leases are hereditary and considered to be permanent. According to the Agricultural Settlement Law of 1967 (Restrictions on Use of Agricultural Land and Waters), non-Jews are prohibited from leasing state land (Haider 1995).

Israel is well known as a regional military power. The Israeli Defence Forces (IDF) are, in fact, one of the principal bodies controlling land in Israel. They regulate large portions of state land, especially that which is mostly inaccessible to civilians. For instance, southern Israel – the Negev – comprises over half of the total area of the state of Israel. It is overwhelmingly state land controlled by the military and only 37 per cent of this area is available for non-military use (Zafrir 1995).

The main objective of local development planning is to implement the national goal of the settlement of the country by Jews (Czamanski and Meyer-Brodnitz 1987). Since the establishment of Israel in 1948, local development plans have been focused on the distribution and dispersal of the Jewish population from the coastal area to other parts of the country. The balance of the Jewish and the Arab populations in certain areas, such as Galilee and the Little Triangle, and the Judaization of all parts of the country have been some of the main goals of the Israeli settlement policies (Yiftachel 1991). In the first fifteen years of statehood, more than thirty new towns were built, especially in the northern and southern parts of the country (Efrat 1994). The defence of the borders and other security issues have also been among the principal concerns of the regional plans. After the 1967 war and the occupation of the West Bank and the Gaza Strip, the creation of new towns shifted to the Occupied

Territories. From 1967 to 1987, the construction of new towns almost stopped within the pre-1967 borders of Israel. Only more recently, following the wave of the Jewish immigrants from Russia and the peace negotiations with the Arabs, has regional development planning refocused itself within the pre-1967 borders.

Administratively, the area of Israel is divided hierarchically into six regions or districts: the North, Haifa, Central, Tel-Aviv, Jerusalem and the South. The districts are further divided into fifteen subdistricts at an intermediate level. The lower tier is the local administrations, which include three types: municipalities (large towns), local councils (small towns) and regional councils (an agglomeration of rural settlements).

By the end of 1998, there were 265 local administrations in Israel. This included 63 municipalities, 147 local councils, 53 regional councils and 2 industrial councils (Israel Bureau of Statistics 2000b). The uneven division of Israel left the six different districts with different loads, especially regarding the provision of services. In the same year, there were seventy-five Arab local authorities: nine municipalities and sixty-six local councils. However, there are still more than seventy unrecognized small Arab settlements without any municipal status. They do not receive services from the central government. The number of Arab local governments with city status is very limited. Although many of the towns are large in size, they are not guaranteed a status of municipality; instead, they have to accept being a local council. The Interior Ministry is often reluctant to give municipality status to Arab local governments, provoking the accusation from Arab leaders that this is because they would be able to demand bigger budgets, especially for development, as well as gaining greater autonomy.

Both Jewish and Arab local administrations share specific legal, political and administrative frameworks. However, with regard to the policies that are carried out, the state and state agencies have dealt differently with Arab local authorities (Falah 1992; Khamaissi 1994). With the appropriation of local authority in Israel, the central government also defined the scope of authority of the local government. For many of the Arab local governments, the central government has not included all the land allocated to the local population in the Arab towns. Part of this land has been placed within the boundaries of adjacent Jewish towns. Hence, Arab landowners have sometimes paid taxes to Jewish local authorities (Falah 1992; Khamaissi 1994).

In addition, local governments in Israel have been granted authority to use the land within their jurisdiction for developmental purposes. They have been authorized to confiscate up to 40 per cent of the remaining private land for public use without paying any compensation to the owners. In Israel, the owners of private land cannot change the use of land without permission from the local authority. The Israeli central government, through gerrymandering and the demarcation of certain boundaries of jurisdiction, transferred resources from Arab to Jewish local governments. By doing so, it blocked the physical growth and expansion of many Arab settlements (Haider 1995; Khamaissi 1994).

Furthermore, if the Interior Minister declares the establishment of a new local administration, he is entitled to appoint the mayor and the local council members. On 24 December 1992, the Director-General of the Interior Ministry, A'mram Qala'ji, as the representative of the Interior Minister, declared the establishment of the Nahal Iron Regional Council, an agglomeration of eight small Arab villages in the northern part of the Little Triangle. The announcement included the appointment of a mayor and five council members. However, all of the members were Jews and none of them lived in the Arab settlements. The name of the regional council is Nahal Iron, a Hebraization of the Arab name of the geographic area: Wadi A'ra. In addition, the offices of this new regional council were set up in Hadera, a Jewish city more than fifty kilometres away from some of the eight villages the council represents.

The map of the jurisdiction of the eight villages that accompanied the declaration did not include most of the land belonging to the inhabitants of the villages. The jurisdictional map of the new regional council was fragmented and the area was quite discontinuous. All of the cultivated lands belonging to the Arab residents were detached and placed under the jurisdiction of neighbouring Jewish regional and local councils.

Following both the objection of the local Arab communities in the eight Arab villages to this appointment of a new regional council and the ruling of the High Court of Justice, the Interior Minister divided the area into two different local administrations in 1997. He set a date – 2 April 2000 – for free local elections for the two different local councils.

The Little Triangle as a region

The 'Little Triangle' is a relatively new term that has become part of the geographical and geopolitical lexicon following the establishment of the state of Israel in 1948. Prior to 1948, the whole area between the Palestinian cities of Nablus, Tulkarm and Jenin was known as 'the Triangle'. The area was primarily connected with the major urban centre of Nablus.

The study area of the Little Triangle is a strip of twenty-seven Arab towns and villages extending along the west side of the 'Green Line', which is the border established between Israel and the West Bank in 1949 (Map 4.2). The rural area covers about 250 square kilometres and extends 50 kilometres from Kafar Qasem in the south to Zalafeh in the north. However, its width ranges from 2 to 5 kilometres. The Little Triangle is territorially continuous, but this continuity has been punctured by a number of Jewish frontier settlements (Shmuali 1985). The region is a rather uniform one. It has a high concentration of Muslim Arabs, who have strong kinship ties with the Palestinian Arabs on the other side of the Green Line. The population, which was 30,000 in 1949, exceeded 170,000 by 1998 (Israel Bureau of Statistics 2000b).

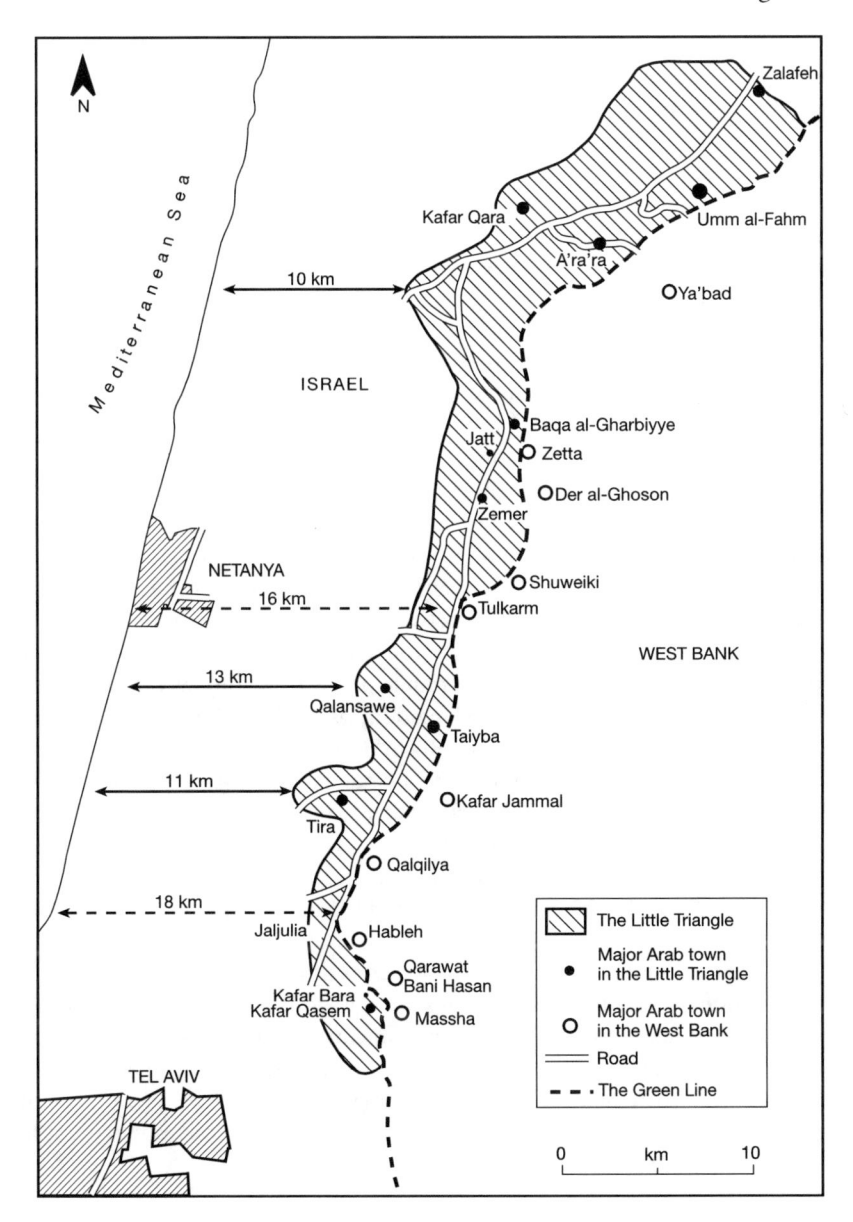

Map 4.2 The Little Triangle and its geopolitical significance

Geopolitical significance

The elongated shape of the Little Triangle along the Green Line enhances the importance of the area. The region is located in one of the most sensitive areas from a geopolitical point of view. The 1949–67 border with Jordan (the Green Line) was very close to the metropolitan area of Tel Aviv and other major Israeli urban centres. The state of Israel has a very narrow 'waist'. The distance between the Green Line and the Mediterranean Sea in the Little Triangle region ranges from sixteen to twenty-six kilometres (Map 4.2). It is crucial to understand the spatial processes that have taken place in this border area since 1949.

Demographically, the Little Triangle continued to be an Arab region with an overwhelmingly Arab population. By 1998, only 43,000 Jews resided in eighteen different settlements in the Little Triangle (Israel Bureau of Statistics 2000b). However, the government of Israel, which controls most of the land, initiated some plans to create a Jewish majority. Several Jewish settlements were built in the region from 1990 to 1996. The creation of Jewish settlements in the Little Triangle since 1948 has led to the fragmentation of the region and the creation of four non-contiguous Arab territorial concentrations: the Kafar Qasem Subregion, the Taiyba Subregion, the Baqa al-Gharbiyye Subregion and Wadi A'ra (Map 4.3).

Although geographically it is located in central Israel, the Little Triangle is considered a peripheral and frontier region. Because it is situated along the Green Line, local governments are unable to expand to the east. They are competing over territories within or to the west of the triangle.

Despite the fact that over 60 per cent of the total area of the Little Triangle is controlled by the state, jurisdictional boundary disputes between Arab local governments and the Jewish Regional Councils still exist. The most persistent disputes are the ones between the municipality of Umm al-Fahm and the regional councils of Menashe and Megiddo, as well as the dispute between Kafar Qasem and the municipality of Rosh HaAyin. On the other hand, many jurisdictional problems have been solved by compromise: for instance, the case of the Menashe Regional Council versus Kafar Qara and Baqa al-Gharbiyye.

The principal jurisdiction of all-Arab towns was decided during the 1950s and early 1960s. During this time, the Little Triangle was under military rule and all the jurisdictions of the Arab towns were decided in consultation with the military governor and the Ministry of Interior, without any consultation with the local residents. Boundary commissions now consult with locals in most cases.

Jurisdictional boundary commissions were established in the 1980s to deal with the grievances of Arab local governments, after which the Arab local governments began to file claims to enlarge their jurisdictions. The establishment of Jewish settlements in the region actually enhanced this process. In the last six years, all-Arab local governments have submitted at least one request for enlarging their jurisdictions.

The questionnaire

Besides interviewing the Arab mayors and heads of all the local governments in the Little Triangle, the author distributed a questionnaire among Arab representatives and professionals in all of the twenty-seven Arab towns and villages in the study area. The Arab representatives were councillors or members of local governments, while the professionals who have dealt extensively with the problems of the Little Triangle included civil engineers, lawyers, architects and

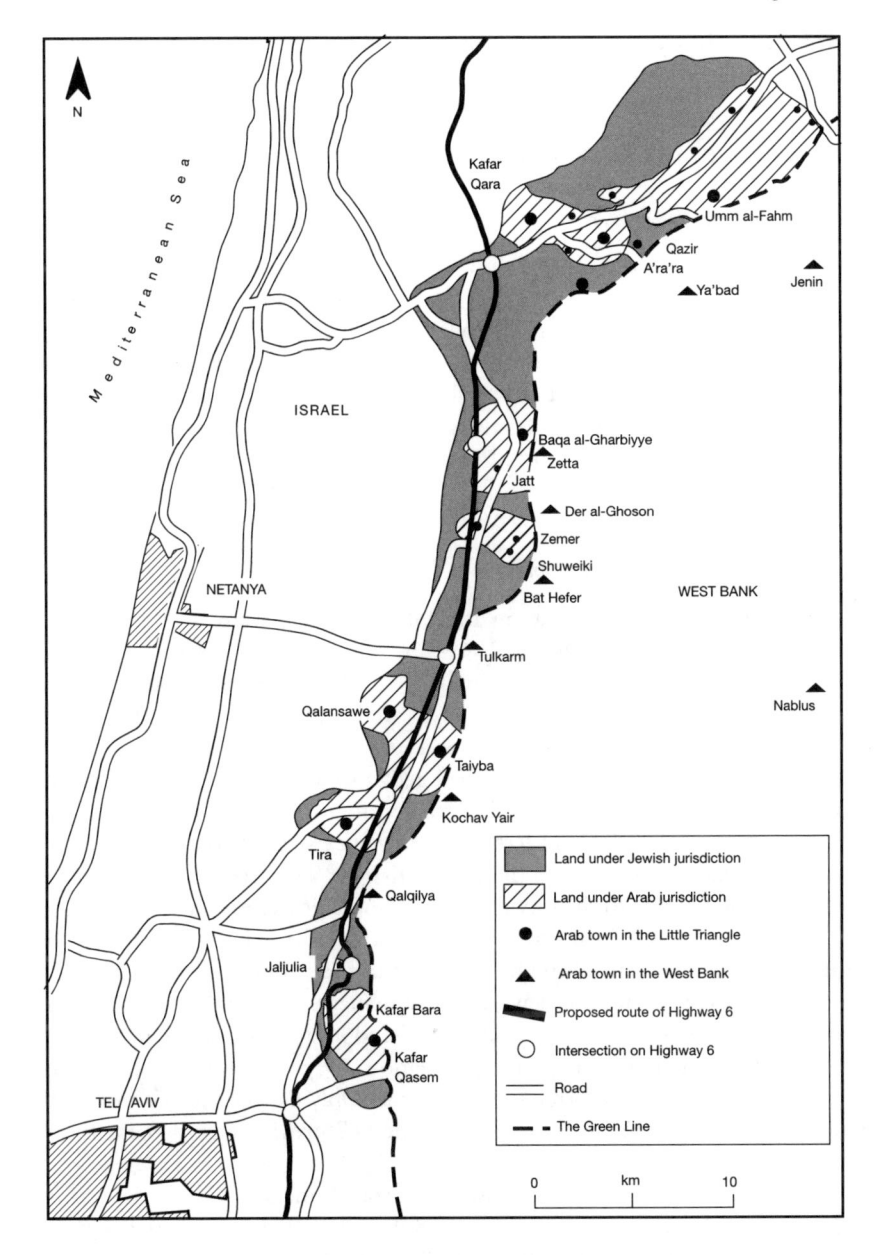

Map 4.3 Arab and Jewish jurisdictions in the Little Triangle

developers. The number of questionnaires, which were distributed during February 1996, depended on the municipal status and the population size of each town. More questionnaires were allocated to large towns with municipal status such as Taiyba and Umm al-Fahm. Out of the 263 questionnaires distributed, 140 were to professionals and 123 to representatives; the response rate was 84.4 per cent.

The results of the attitude survey

The decision-making process

The results of the questionnaire show that 85 per cent of the Arab representatives believe that the decision-making process is unfair and not acceptable, while 92 per cent believe that the Israelis have favoured political considerations rather than economic or social ones as the most important factor in determining the jurisdiction of Arab towns. Furthermore, only 39 per cent believe that local Arab claims have any influence on the government's decisions. The heads of all-Arab local governments supported these results and the mayors showed little trust in the decision-makers.

The creation of Jewish settlements

Most of the Jewish settlements in the Little Triangle were built in the early years of statehood, mainly during 1949–55. These settlements were agriculturally oriented and few people inhabited them. They were considered frontier settlements, serving a security function as well. However, within the last few years, new development occurred in the Little Triangle in the form of accelerated Jewish construction activities. These activities were not on the basis of pure market demands, but were due to the intervention of the Israeli authority.

The survey of local Arab leaders shows that they strongly oppose any further Jewish construction activities in the Little Triangle. They regard any new Jewish settlements as a threat to their own existence and culture. Among the responses, 82 per cent acknowledged that they looked on the construction of Jewish settlements as a threat to Arab physical space in the region. Arab leaders believe that the economic benefit of new Jewish settlements for the Arab population is negligible: While only 16 per cent considered the Jewish settlements to be economically beneficial to the Arab population, 69 per cent rejected this statement. This is also true regarding employment. Only 22 per cent claim that the new Jewish settlements in the Little Triangle will provide employment for the Arab population.

According to the responses, most local Arab leaders of the Little Triangle (93 per cent) believe that the creation of Jewish settlements in the region is largely influenced by the existence of Arab towns close to the Green Line along the West Bank. In addition, the inhabitants of the region see a connection between Jewish construction activities and the arrival of hundreds of thousands of Jewish immigrants (88 per cent). On the other hand, most of the surveyed people (60 per cent) do not see any connection between the creation of new Jewish settlements in the region and the possible eviction of Jewish settlers from the Occupied Territories.

Finally, one positive outcome of creating new Jewish settlements in the Little Triangle, as seen from the responses, is that they encourage Arab local governments to file claims requesting the enlargement of Arab local government jurisdictions. Among the responses, 65 per cent supported this statement, which was also supported by most of the Arab mayors and heads of local governments during personal interviews.

Jewish–Arab relations in the Little Triangle

Most of the people surveyed support common projects between Jewish and Arab local governments (81 per cent). The Arabs, in the region studied, strongly feel that they are not welcome to live in the new Jewish towns in the area (96 per cent). In addition, Arab leaders prefer to have common projects with other Arab local governments (94 per cent). However, the majority (55 per cent) rejects any incorporation with other Arab local governments. Many believe that the incorporation of Arab local governments will only lead to the reduction of the jurisdiction of Arab towns. Most of the people surveyed mentioned the outcome of the incorporation of the towns of A'ra and A'ra'ra to one local government – and the eight tiny villages of Nahal Iron to one regional council – where the towns lost most of their agricultural area to the adjacent Jewish regional council.

Judging from the in-depth interviews with the heads of the Arab local governments and the heads of Jewish regional councils in the Little Triangle, cooperation between them is quite limited. There are, however, several common projects in the area. These include a sewage treatment plant, which the Arab local governments of Kafar Bara and Jaljulia share with the Jewish Derom Hasharon Regional Council, and another sewage treatment plant common to Qalansawe and the Lev Hasharon Regional Council. There are three more shared sewage treatment plants planned or under construction in the area: plants between the municipality of Tira and the Derom Hasharon Regional Council, between the Wadi A'ra Arab towns and the Menashe Regional Council, and between Zemer and the Emeq Hefer Regional Council.

In addition, the Arab and Jewish local authorities in the study area share several regional services. These services are provided by the central government and include environmental protection units, police and fire stations, as well as rubbish dump locations.

The mode for cooperation between Arabs and Jews in the Little Triangle is an advanced one, following the larger peace process. Cooperation with Palestinian municipalities on the East Side of the Green Line in the West Bank has increased as well. In the study area, several industrial parks are now planned, including an industrial park common to Jaljulia and the Derom Hasharon Regional Council, Taiyba and the Lev Hasharon Regional Council, and Baqa al-Gharbiyye and the Menashe Regional Council. Another industrial park will be shared between the Emeq Hefer Regional Council and the Palestinian municipality of Tul-Karm.

In addition, all the heads of the Jewish regional councils and those of the Arab local governments reject the idea of any type of municipal incorporation. While the Jewish representatives claim that the type of development and cultural differences make the task of municipal incorporation impossible, the Arab representatives assert their desire to keep their land private for themselves. They add that the central government has been using the incorporation of Arab local governments as a tool to reduce the jurisdiction of Arab local governments and to control, if not confiscate, Arab land. Most Arab heads reject the idea of municipal incorporation of Arab towns as well. They often quote the incorporation of A'ra and A'ra'ra as examples of the loss of Arab land.

The disparities between the quality of life in Arab and Jewish towns in the study area have encouraged several Arab families to try moving to the Jewish towns. Arab leaders expect this trend to increase, especially among some of the Arab youth, if the living conditions in the Arab towns do not improve. The Arab population yearns to benefit from the high standards of living in the newly established Jewish towns. This trend is strongly rejected by the four Jewish regional council heads that I interviewed. Nachum Itzkovitz, the head of the Emeq Hefer Regional Council, for instance, was in favour of planning mixed towns for Arabs and Jews, but he opposed the housing of Arabs in already existing Jewish settlements within the region.

Another dimension of this study was to examine the decision-making process in Israel regarding the issue of allocating land to Arab local governments, the impact of these policies on the Arab population in the study area and how it was perceived by the local population. The findings of the study support the hypothesis that the decision-making process is unfair and discriminatory. As became clear from the first part of the questionnaire, individuals representing the Ministry of the Interior during the military government regime in the 1950s and early 1960s defined the jurisdictions of most of the Arab towns. The Arab leaders and representatives perceive the decision-making process regarding the area of land allocated to local government in Israel to be very difficult and not efficient, as well as being discriminatory in the case of Arab local governments. While the process of any jurisdictional boundary enlargement request takes many years for the Arab local governments, it takes much less time in the case of Jewish municipalities.

As mentioned previously, when the Arab local governments were established in the 1950s and 1960s, the Ministry of the Interior determined the jurisdictional boundaries of each town. No one consulted the local residents in the Arab towns. However, this rather arbitrary process no longer applies. Since the early 1980s, jurisdictional commissions have been established. In most cases, an Arab representative has been included in the commission dealing with Arab towns. However, some Arab leaders quote the case of the Arab Nahal Iron Regional Council, which was established in 1994, to prove that not much has changed during the 1990s. In this case, the local residents were consulted only after the decision to establish this council was made by the Interior Ministry.

The fact that jurisdictional commissions are limited to only making recommendations, leaving the ultimate decision in the hands of the Minister of the Interior, is another concern for the Arab leaders. For most of the time since the establishment of Israel, religious parties have controlled the Interior Ministry. Issues related to Arab land and land in general are treated differently by the Interior Ministry, according to several local Arab government heads. Although both Arab and Jewish local governments can and have petitioned the High Court of Justice against the Minister's decisions, Arab leaders believe that the jurisdictional boundaries of a town should be decided by professionals rather than political leaders.

In addition, the reduction of Arab land and Arab local government jurisdiction led to the reduction of land for housing and infrastructure. Land available for housing is a scarce commodity in Israel, especially in the Arab towns and villages of the Little Triangle. Land prices in the Arab towns are very high. An average *dunum* of land costs US$80,000–140,000. This is in contrast to average cost of US$20,000–30,000 per *dunum* in Jewish towns of the Little Triangle such as Qazir, Bat Hefer and Cochav Yair (Map 4.3). It should be noted, however, that the cost in the Arab towns is for ownership of the land, while in the Jewish towns it is for the right to lease the state land for forty-nine years, after which the lease may be extended. Moreover, the high price of land for housing has led the people to move from the traditional one-storey house to buildings with three to five storeys.

The analysis of the reactions of the Arab leaders to the Israeli decision-making process clearly shows their frustration and self-assertion. There is a growing level of local mobilization as has been seen in the case of the Nahal Iron Regional Council, as well as increasing demands for changing the decision-making process towards increased participation of local residents. The central government treats the local Arab government as totally dependent agents, who must accept whatever land area is allocated to them.

According to the respondents, the Arabs perceive Jewish settlement plans in the region as a threat to their physical space. Any radical transformation of the physical and cultural landscapes in the region is found to be counterproductive. The continuing Judaization policy and its attempt to bring about a Jewish majority by the year 2020 will lead to rising tension and threaten the social order and political stability. In contradiction of the research of some scholars, the Arabs of the Little Triangle enjoy minimal economic benefits from the creation of these settlements. In addition, the respondents believe that the construction activities in the region are connected to the arrival of hundreds of thousands of Jewish immigrants from the former Soviet Union.

As the last part of the survey shows, this has had an impact on Jewish–Arab relations. The Arabs have a complex relationship with the Jews. In spite of the fact that the relationship between the two ethnic groups cannot be considered good, the Arab leaders demand common projects with Jewish local governments. This can be explained by the belief that Jewish local governments have better access to central government agencies. Through common projects,

Arab local governments can maximize their development. Likewise, most of the Arab leaders strongly support common projects and cooperation with other Arab local government. This attitude can be explained by the need to maintain their national unity in opposition to the dealings of the central government with them, as individual Arab towns and different ethno-religious groups within Israel.

Equally important, public spaces in the Little Triangle where Arabs and Jews can or are forced to meet are very limited. Besides the fact that many Arab workers commute to the metropolitan Jewish areas each day, Arabs and Jews meet during weekends in football stadiums and a limited number of Arab markets. However, such meetings have been dramatically reduced following the violence that erupted in the Arab towns in October 2000 between the Arab citizens and the Israeli police.

Finally, based on the findings of the study, Arabs as well as Jews prefer the development of their towns without an ethno-spatial mixture. However, the conditions in the Arab towns, especially with regard to the issue of land prices and housing, will force Arab youth to seek residence in adjacent Jewish towns, where they feel strongly unwelcome. The case of the Qi'dan family is an example of what lies ahead.

In April 1995, the Qi'dan family submitted a request to buy a house in Qazir, a new Jewish community in the northern part of the Little Triangle. The land is cheaper and services are far better in Qazir than any Arab town in the Little Triangle. The local government of Tal-Iron, to which the community of Qazir belongs, refused to sell a house or a piece of land to the Arab family because Arabs are not allowed to live in this new town. Qazir was built in 1982 specifically for Jews, as were dozens of other towns and settlements in Israel controlled by the Jewish Agency. The Agency receives land from the Israeli Land Authority (ILA) and is only authorized by the central government to sell homes and lease land to Jews. The Qi'dan family also demanded the lease of a piece of land directly from the ILA, but their request was again rejected.

In July 1996, the Qi'dan family, through the Israeli Association of Citizens' Rights, a well-regarded Israeli organization, petitioned the High Court of Justice asking for an explanation. The family requested a ruling against the Jewish Agency, the ILA and the decision of the Tal-Iron local government, for not allowing them to live in Qazir. In the petition, the attorney who represented the Arab family emphasized the discriminatory policy against her client:

> This discrimination contradicts the essence of Israel as a democratic country ... [and] based on the basic law of man's dignity and freedom, the state should balance between being a Jewish and a democratic state.
>
> (*al-Itihad* 19 March 1996: 5)

In their response to the petition, the representatives of the Jewish Agency (which is funded internationally) emphasized the fact that it is only authorized to build settlements for Jews. The Agency sees itself as playing a vital role in achieving the national goal of settling Jews in Israel. In its response, the Tal-Iron local government also stressed the goal of the establishment of Qazir as part of an effort to:

> increase the Jewish presence in the Little Triangle. Absorbing Arabs in Qazir will threaten the goal of its establishment and will lead to many Arabs who will follow.
>
> (*al-Itihad* 19 March 1996: 5)

Finally, on 3 March 2000, the Israeli High Court of Justice (HCJ) ruled that discrimination against Israeli Arab citizens in the allocation of state lands is illegal. The experts saw the ruling, in what has come to be known as the 'Qazir Case', as a landmark decision that will influence the future planning and geography of the country. However, in December 2000, the Qi'dan family petitioned the HCJ again, this time for contempt of court against the ILA, because the ILA had not carried out the court decision to allow the Qi'dan family to move to Qazir.

Conclusion and theoretical implications

In this study, I argue that land in Israel has more than merely economic significance. It conveys cultural and political meanings that have been motivating both the Arabs and Jews. The Arab citizens of the Little Triangle have lost most of their land due to the new political realities that followed the establishment of the state of Israel. The 1948–9 war and its outcome, as well as the massive Jewish migration to Israel, led to large-scale confiscation of private Arab land. Over the years, the uniform Arab region of the Little Triangle has lost its character and witnessed considerable Judaization.

The fragmentation of the Little Triangle started soon after its incorporation into the new state of Israel. As mentioned previously, from in the early 1950s a series of laws and regulations provided the legal means for a large-scale expropriation of land from Arab ownership and its transfer to Jewish hands. This process continued throughout the 1960s and 1970s and the outcome has had a tremendous political impact on the Arabs and their land. The uniform territory of the Arab Little Triangle soon became fragmented, with the majority of the land becoming Jewish. During this period the establishment of most of the Arab local governments and their jurisdictions in the Little Triangle was also determined, even though there have been slight changes in jurisdictions over the decades that followed. Today, the Little Triangle region does not function as one unit, but is divided into four distinct Arab subregions: Kafar Qasem, Taiyba, Baqa al-Gharbiyye and Wadi A'ra (Map 4.3).

The present analysis concentrates, also, on the impact of Judaization on the Arab towns and population in the Little Triangle. The survey of the attitudes of the Arab leaders and representatives reveals that the Arab population in the study area feel that their physical space and existence in the region is threatened.

This feeling is mainly an outcome of the reduction of Arab land ownership, accelerated Jewish construction activities and the fact that this predominantly Arab region is projected to have a Jewish majority by the year 2020.

The Arab leaders and representatives believe that the decision-making process regarding the jurisdiction of towns is unfair and discriminatory. Set up in such a way as to reduce Arab land ownership, this process is impeding the development of Arab towns. The Arab leaders believe that economic and social factors, rather than political ones, should determine the jurisdiction of towns. The Arab towns should not be at the mercy of the Minister of the Interior, who believes that land should not be sold to non-Jews (as stated by one Arab leader in the region studied).

The findings of the study refute the claim, often expressed by Israeli decision-makers, that Arabs and Jews get equal treatment on the issue of jurisdiction. Furthermore, it contrasts with the assertion of equal economic benefits, especially with regard to employment in the Little Triangle. Although the Arab population has improved its economic situation, considerable economic disparities between Arabs and Jews continue to exist. Following occupational and income inequalities and discriminatory allocation of resources to the various sectors within Israel, particularly in the Little Triangle, the socioeconomic status of the Arabs in the region studied is rather lower than that of the Jewish Israelis.

The Arab population supports cooperation between Arab and Jewish local governments. The lack of basic amenities in the Arab towns and better economic development in the Jewish towns encourages Arab leaders to support and sometimes even initiate common projects. Many believe that, through cooperation and common projects with Jewish local authorities, the Arab towns can achieve greater development. However, there is more support for projects between Arab towns. This may be explained by the desperate need for better development in all-Arab towns.

A report, written by twenty-five experts and researchers and submitted to the Israeli prime minister following the disturbing events in the Arab towns of Israel during October 2000, stated that:

> While the Arab population continues to grow, the amount of land available to it dwindles as a result of appropriation and annexation of Arab land by Jewish councils. Land and planning policies consistently discriminate against Arabs. Virtually no state land is allocated to existing Arab towns and villages; new Arab settlements are not established with state help; Arab access to state-owned land is blocked-off ... . Land policy and planning in Israel is based on patterns of discrimination, which are at variance with basic principles of equality and democratic administration. The current situation can be rectified only if perceptual and structural changes pertaining to the land/planning sphere are implemented. Democratic, egalitarian policies regarding land rights and policy enforcement must be adopted. Such change would reinforce belief in the legitimacy of the governmental system

among the Arab population and it would contribute to Israeli's political and social stability.

(Rabinovits, Ghanem and Yiftachel 2000: 13)

On the theoretical front, the literature on planning and economic development does not discuss in detail the mechanisms of land allocation. Land is the most important commodity and it is presumed that there is always land for trade. The literature does not go beyond the economic value of the land and property. However, what is driving the Arabs and Jews with regard to the land issue is not merely the economic but also the political and cultural gains. This important issue is absent in theories dealing with Western cities.

Therefore, what is needed is a theory that focuses on the context of rapid economic development, that examines local government in relation to the historical development of the state, in a setting of ethnocratic ideals and ethnic segregation, and deals with conflicts based on the distribution of resources – namely, land.

References

Cadwallader, Martin (1996) *Urban Geography: An analytical approach*, Upper Saddle: Prentice Hall.
Cohen, Amnon (1965) *Arab Border Villages in Israel*, Manchester: Manchester University Press.
Czamanski, Daniel and Meyer-Brodnitz, Michael (1987) 'Industrialisation in Arab villages in Israel', in Raphael Bar-El and Arieala Nesher (eds), *Rural Industrialisation in Israel*, Boulder: Westview Press.
Efrat, Elish (1994) 'New development towns in Israel', *Cities* 11(4): 146–53.
Falah, Gazi (1992) 'Land fragmentation and spatial control in Nazareth metropolitan area', *The Professional Geographer* 44: 30–44.
Geertz, Clifford (1963) *Old Society and New States*, New York: Free Press.
Ghanem, Asa'd (1993) *The Arabs in Israel towards the 21st Century: A study of basic infrastructure*, Giva't Habiba: Institute for Peace Research.
Haider, Aziz (1995) *On the Margins: The Arab population in the Israeli economy*, New York: St Martin's Press.
Israel Bureau of Statistics (2000a) *Statistical Abstract of Israel 2000*, Jerusalem: Central Bureau of Statistics, no. 51.
—— (2000b) *Israeli Statistical Annual: Local authorities in Israel in 1998: Physical data*, Jerusalem: Bureau of Statistics, no. 1134.
Keeble, David (1976) *Industrial Location and Planning in the United Kingdom*, London: Methuen.
Khamaissi, Rasem (1994) *Towards Strengthened Local Governments in Arab Localities in Israel*, Jerusalem: Floersheimer Institute for Policy Studies (in Hebrew).
Lustick, Aan (1980) *The Arabs in the Jewish State: Israeli control of national minority*, Austin: University of Texas Press.
McCrone, Gavin (1969) *Regional Policy in Britain*, London: Allen & Unwin.
Rabinovits, Dani, Ghanem, Asa'd and Yiftachel, Oren (2000) *After the Calamity: New directions to governmental policy toward the Arabs in Israel*, Jerusalem: Shatil (in Hebrew).

Ross, J. Allen (1982) 'Urban development and the politics of ethnicity: a conceptual approach', *Ethnic and Racial Studies* 5: 440–56.
Schofield, John (1987) *Cost–Benefit Analysis in Urban and Regional Planning*, Boston: Allen & Unwin.
Scott, Allen (1980) *The Urban Land Nexus and the State*, London: Pion.
Shmuali, Amir (1985) *The Little Triangle: Transformation of a region*, Haifa: Institute of Middle Eastern Studies of Haifa University (in Hebrew).
Soffer, Arnon (1983) 'The changing situation of majority and minority and its spatial expression: the Arab minority in Israel', in Nurit Kliot and Sally Waterman (eds), *Pluralism and Political Geography*, London: Croom Helm.
Yiftachel, Oren (1991) 'State policies, land control, and ethnic minority: the Arabs in the Galilee Region in Israel', *Society and Space* 9: 329–62.
—— (1995) 'The dark side of modernism: planning as control of an ethnic minority', in Sophie Watson and Katherine Gibson (eds), *Postmodern Cities and Spaces*, Oxford: Blackwell.
Zafrir, Rinat (1995) 'The deer and the tank country', *Haa'ritz* 28 July: 17 (in Hebrew).

5 Borders and boundaries in post-war Beirut

Daniel Genberg

Introduction

In her review of anthropological writing on cities, Setha Low (1996) lists a number of metaphors or images that correspond to certain concepts and frameworks in approaching the city. Two of these are 'the ethnic city' and 'the divided city'. One stream of research within the former is 'studies of the ethnic city as a mosaic of enclaves economically, linguistically and socially self-contained as a strategy of political and economic survival' (Low 1996: 388). The boundaries containing these enclaves are, more often than not, changing, flexible and permeable. On the other hand, the latter – the divided city – points to a city where segregation has divided the city in two (most often unequal) parts. Crossing the border that divides this kind of city is often difficult, since segregation is often maintained by one of the sides (upscale versus ghetto, black versus white, poor versus rich etc.).

Both of these images have some bearing on the city of Beirut. It was to a large extent a divided city during the civil wars. This division was not perhaps desired by either side and was due more to the logic of warfare, but it was not entirely arbitrary and certainly not without consequences.

Beirut cannot qualify, in practice, as an ethnically diverse city. The differences projected by, or on, the various groups of people are in most cases not formulated in terms of ethnicity but in terms of religious or sectarian affiliation. These differences have not always been decisive in the choice of habitation; they have, in certain circumstances, been of critical importance. More importantly, the various neighbourhoods have been conceived by people as being dominated by one group or another, or being under the control of a certain militia during the war. Very few neighbourhoods are entirely homogenous and very few areas are completely mixed. The aim of this chapter is not to geographically map the religious and sectarian diversity of Beirut but rather to try to give an account of the perception and experiences of its inhabitants in relation to urban space in post-war Beirut.

There seems to be one theme in practically all the research which has been done on cities that has remained unchanged and has become central for the chapters in this volume. It is usually expressed with such terms as 'diversity', 'difference', 'variety', 'multitude', 'heterogeneity' and so on. For a human settlement to be called 'city' it needs to be 'a relatively large, dense and permanent settlement of socially heterogeneous individuals' (Wirth 1938: 8). The sameness and familiarity of village life is replaced by its opposite, something that amounts to the image of cosmopolitan life.

Today, the pace of urbanization has many times surpassed the 'alarming' growth of cities 150 years ago. Segregation, in the form of the continuing growth of ghettos – in which minority ethnic groups are forced to live as a result of discrimination – and the emergence of gated communities – into which communities willingly 'gate' themselves for protection – show no signs of disappearing. This is one of the reasons for the heightened attention paid by both international organizations and national institutions to the future of the city.[1]

One question that emerges is what the relationship is between cosmopolitanism and segregation. A segregated city, seen as a whole, would be diversified and heterogeneous, thus cosmopolitan or urban. The problem here lies in the perspective we take. Once the various communities – whether in terms of ethnicity or class, since the two are often linked – fend off each other spatially we could be said to be one step removed from the 'liberating' or 'overwhelmingly' urban atmosphere[2] that stood as a central pillar for early urban analysts. What then remains of the city, once it is split into 'villages' without much interaction? It is at this point that it becomes vital to look more closely at both the boundaries and the spatial aspects of any encounters. This is why I have turned to the case of the city of Beirut, which I hope will show that borders and boundaries change over time and depending on where one is situated.

Beirut in war and peace

This chapter is not about the war itself, but it is difficult to pass on to the recent developments in Beirut without even briefly mentioning it. In fact, one does not have to give an account of this war, since so much of the literature on Beirut is full of it and the image most people foreign to Beirut have of the city is about the war. Indeed during the civil war period in Lebanon, the term 'Balkanization' was popularly recoined as 'Lebanonization'; the tragedies in the Balkans over the 1990s have led to a revival in use of the original term.

The fighting that ravaged Lebanon and Beirut from 1975 to 1991 was not continuous. What is generally referred to as 'the civil war' should perhaps be expressed in the plural instead: During the sixteen-year slot of time the war has been allocated in history, there were periods of peace, ceasefires and periods of tranquillity. The wars of Tripoli in northern Lebanon were not like the ones in Beirut. The groups and militias involved in the wars were numerous. One thing that is important to keep in mind is that the fighting was not between the communities in the country but between militias. Some of these waved the banner of a certain community, but that is not to say that all in that community were behind that militia. Indeed, a great number of the clashes were between militias claiming their turf within the same community. Some militias were created out of a political adherence to an ideology that was, in part, anti-sectarian. Other actors in the wars were armies from other countries, notably those connected with the Israeli invasion in the early 1980s.

The political system of Lebanon is sectarian or denominational. That is, parliamentary seats are allotted to the elected according to religious identity. The three top political positions – President, Prime Minister and Speaker of Parliament – are reserved for a Maronite Christian, a Sunni Muslim and a Shia Muslim respectively. This has been the case since the creation of the state of Lebanon in 1943. The Constitution of the Lebanese Republic had been drafted in 1926, during the French Mandate (1920–1943), and was accompanied by a sort of backroom deal (Makdisi 1996) or gentleman's agreement (Salibi 1988: 185), known as 'the National Pact', which ensured the power-sharing between these different groups.

This does not mean that all of its citizens agree with this system or even that everyone finds it of primary importance to socialize only within one's own circle. Sectarian affiliation used to be printed on the identity card, something abolished a few years ago, but the identity of someone can still be rapidly established by their name, place of birth, place of birth of their parents and, to a certain extent, by their accent. Some family names are recognized as the names of important families within a community. Also, there is a difference between being called 'Joseph' and 'Youssef'. One young man I encountered had been given the first name 'Charles de Gaulle' by his parents, leaving little room for doubt as to his sectarian background, although he presented himself as 'Charles' only. Clothing, beards, haircuts and other adornments are also used to identify people according to their affiliation. This classification is often immediate and almost unconscious, having no further consequence after having been acknowledged. On one occasion during my fieldwork, I was looking for an apartment together with four of my informants. The ones who were to sign the lease were married. As the owner of the apartment was asking our names she had no trouble spotting me as a foreigner. The married woman posed a problem. Her first name, maiden name and family name did not give her away. Nor did her place of residence, in the Chouf mountains, since it had been in the hands of both Christians and Druze – likewise, her place of birth. Her mother's maiden name was Russian and did not make things any clearer. Only on asking the place of birth of her father and the maiden name of her grandmother on her father's side did the landlady determine that the wife of her prospective tenant was Druze. But in the end this was of no consequence, since the landlady then said:'Welcome!' It seems that she only wanted to know out of curiosity and from habit.

In spite of people being sometimes hard and at other times easy to identify in terms of sectarian background, villages and neighbourhoods are for the most part very easy to pinpoint. Again, the names of neighbourhoods are, for anyone living in Beirut, directly associated with a community. Flags, posters, shrines and graffiti adorn most streets and neighbourhoods. A Virgin Mary inside a glass box; 'Jesus of Nazareth, King of Kings' sprayed on the wall; the green-black-red-and-white symbol of Amal (a Shiite militia/party); the yellow flag of Hizballah; the red symbol of the Syrian Nationalist Party; a Palestinian flag; a Syrian flag; a Lebanese flag; the Christmas lighting; the symbol of the Kataaib party; a photo of Hariri or Jumblatt or Berri or Gemayel; the names of shops; the ringing of church bells or a call to prayer ... they all function as markers of territory and tell more than one story. They do not necessarily function as borders keeping people out, or as gates keeping people in, but they do signal to anyone coming into the neighbourhood who is or has been in control of the area.

Dividing history

As can be understood, there is a system that divides the political entity of Lebanon into communities. It is also something that exists in everyday reality, even if many people say they do not attach any importance to religious identity. In pre-war Beirut, different neighbourhoods were known to be of a certain sectarian majority. However, there was no policy that prohibited others from living in a particular neighbourhood, hence, totally homogenous neighbourhoods did not exist.

The first census of Lebanon dates to 1932, when Lebanon was a state under French Mandate. People were counted and grouped according to community affiliation. We learn that the population of Beirut was 179,370, the largest groups being the Sunni Muslims followed by the Maronites and then the Greek Orthodox. The number of foreigners (meaning those from France, Italy, Greece and so on) at the time was 14,554. The early twentieth century had seen large waves of migration to the city. For example, in 1922 around 20,000 Armenians had come as refugees and been assigned by the French authorities to live in the quarantine area just northeast of the city. Other migrants to the city followed this pattern of settling according to identity. The old city centre had known a similar division of quarters, or gates, each bearing the names of a prominent family, but the period of the French Mandate did see a 'consolidation and sectarian "homogenization" of [the] peri-urban quarters' (Saliba 1998: 13).[3]

Faced with the rapid growth of population and influx of migrants from the countryside and neighbouring regions, Beirut came to the point of needing a plan for expansion and reconstruction or modernization. In 1932 'the Danger plan' was presented by French consultants. It included both an analysis of the situation and proposals for change. In it we learn that:

> Beirut's population is extremely diverse and composite. The physiognomy of the quarters indicates well enough the general mixture of races and religions. There are no quarters that are strictly Muslim or Christian with their walls and gates reinforced, like in Aleppo. Here we are in a Mediterranean port where commerce and navigation have always facilitated exchange.
> (Danger 1932: 10–11)

The period of civil strife changed this picture radically, as people moved due to emerging hostilities. This changed the composition of certain quarters. For example, in 1975 the Christians made up 35 per cent of the population in western Beirut. This number was down to 5 per cent in 1989 (Nasr 1993: 69). The eastern suburbs of Beirut had a 40 per cent Muslim population in 1975, but only 5 per cent in 1989. This does not simply mean that Muslims moved west and Christians to the east, but rather that there was a homogenization of certain areas as a result of people moving out of the city or out of the country.

The years preceding the war had seen a rapid growth of urban population across Lebanon, notably in Beirut. In 1959 Greater Beirut's population was around 28 per cent (450,000) of the country's total population; in 1975 migration (from the countryside and from abroad) to Beirut had increased this proportion to almost 50 per cent (1,200,000) (Mouzoune 1999: 14). By 1989 it was estimated that 45 per cent of the population of Lebanon had been displaced on a long-term basis (Nasr 1993: 67), a large number of whom had emigrated. In fact, the figures and numbers vary greatly depending on sources. The number of people living in Beirut changes, depending on whether one takes into account the municipality of Beirut or the urban agglomeration Greater Beirut. For the municipality of Beirut we find, in 1970, around 475,000 inhabitants (Mouzoune 1999: 20) and, in 1996, about 407,000 (Figuié 1997: 35). These numbers include only Lebanese citizens. The proportion of the different religious communities are even harder to estimate as an official census does not exist and approximations vary quite a lot: for example, in 1996 the estimate for the Christian/Muslim ratio varied between 30:70 and 45:55 (Figuié 1997: 38). Not counted in the numbers above are the foreigners residing in Lebanon, mostly from other Arab countries, who numbered around 1,100,000 in an estimate for 1994 (Figuié 1997: 75).

The war changed what could be considered a trend towards pluralism. Instead we witness a homogenization of territory and segregation of the population. This territorial logic was established by the war, not the cause of the war, as:

> each armed faction strove to create its own homogenous sectarian territory through intimidation, threats and violence against political opponents within its community and against members of other communities living within the borders of its sectarian territory.
>
> (Nasr 1993: 68)

The fighting also made the spaces for living smaller:

> With internecine conflict, quarters within urban districts, just like towns and villages, were often splintered into smaller and more compact enclosures. Spaces within which people circulated and interacted shrunk still further.
>
> (Khalaf 1993: 33)

As these areas became more homogenous, as well as becoming more centred on themselves, we also witnessed the disappearance of those areas that had been the centres for so many civic activities and had not been marked by any denomination. For example, soon after the outbreak of hostilities, the central district – with its parliament and municipal buildings, financial district, transportation network, souks, shopping areas and night entertainment areas – was inaccessible to most inhabitants. As the militias fought over control of the city, the city centre of Beirut took on a new geography in the minds of people. Tall buildings became of central importance and also a source of fear in the daily routines of people. For instance, a shop owner on rue Gouraud, east of the central district, told me that out of fear of snipers he would only drive on the northern side of this street, since it was not visible from the tallest building of Beirut, the Murr Tower just west of the centre. Sites of prolonged fighting became very familiar to the people through the continuous repetition of news reports. Barbir, the Khalde intersection and Mathaf became important areas in the geography of Beirut; the route between Mathaf (the National Museum) and Barbir had been a kind of neutral ground during the war and served as a passage between east and west. Right after the war there were plans to construct a new government and parliament complex along this axis. Today these remain connection sites for different taxi routes and bus services. These became the names of borders that were either closed or open for going over to the other side.

The city of Beirut has been parcelled out. Since the city centre became inaccessible, or at least out of reach, a number of other centres emerged which took over many of the functions previously filled by the city centre, creating parallel 'cities' in a shared space of fear. It is important to remember that, although the city was divided at times by lethal lines of confrontation, the constituent parts of the city were still connected. Cherished friends and relatives who lived in other parts of the city or in different places alogether maintained vivid and real connections with the inhabitants. Also, a lot of people did risk crossing dangerous parts of town to do or find what they needed or wanted.[4]

Consolidating boundaries

Part of my fieldwork in Beirut focused on a neighbourhood called Gemmayze, located just east of the downtown area.[5] I had been given the advice, by some residents, to look up a neighbourhood group that organized different events to boost morale and promote the area. After getting the approval of the financial sponsor of this group,[6] I was invited by Amjad, one of the active members, to his house for an interview. He had been active in the Kataaib party (a Maronite militia advocating Lebanese nationalism with a Christian undertone) as the member responsible for contact with the media. Once in his house in Jounieh, a town north of Beirut that had grown exponentially during the war because of people fleeing the fighting in Beirut, he wanted to tell me the history of Gemmayze:

> Gemmayze is Christian. It has always been Christian. Originally it was Orthodox but then Maronites came from the mountains and now they are a majority. It is one of the oldest areas of Beirut. I mean, in Beirut outside the old town. Lebanon is a bird with two wings, one Christian and one Muslim. Beirut is a bird and Gemmayze is one wing and Basta is the other. If one would give an example of a Christian street one would say rue Gouraud [the main street in Gemmayze]. In Beirut there are the boys from Gemmayze and the boys from Basta. They united to form Lebanon and during the war of 1975 they fought each other. Now Gemmayze is one of the oldest areas in Beirut because most of the old city has been destroyed. If one would want to show a typical old street of Beirut one would go to rue Gouraud.

After a while we were interrupted by some friends of Amjad who had come to visit. One of them explained that they had all fought in different militias during the war. Joe explained:

> You know, we were fighting each other and now we are sitting on the same sofa. I was in a communist militia allied to the Palestinians for a while. Then I escaped and went to the USA, got an education. I am an architect now. It is perhaps strange but it was in the States that I learnt to hate the Muslims even though I was on their side when I was here. Now, I sometimes go to Beirut only to work. I prefer to live in Jounieh.

Later that night Amjad drove me home. I was living in Hamra, which is on the other side of downtown from Gemmayze, about two kilometres away. As we passed through the desolate city centre, Amjad's knuckles went white from firmly grasping the steering wheel.

> You have to guide me now. I have never been on this side. I don't need to go here. All I need, I can find on the other side. That is something nice with Beirut; everything you need you can find in your neighbourhood.

This was in 1997, some seven years after the end of the war.

Amjad's Beirut still bore the geography of the war. He is not the only one in this landscape of fear, but it is far from being shared by everyone. One day not long after crossing the city with Amjad, I was taking a walk with another informant who had been studying, for four years, at a university in Beirut. The fighting had been a constant backdrop to his childhood but he had never considered joining any militia, partly due to his young age. For him there was no 'other side of the city'. Downtown Beirut was at the time and still is a huge construction site, displaying empty concrete slabs, glass buildings, renovated houses, bombed-out buildings, new and old roads, and an array of pictures showing what a building will look like once finished. All this was covered by a thin layer of freshly fallen dust. All the construction activities in this area of two square kilometres was supervised by a private construction company, popularly known as SOLIDERE,[7] which was officially launched in 1994. As we entered the 'SOLIDERE area' we left daily life behind; the only people you could see on foot in this area were people working there or tourists and teams of journalists. Others passed through it by car, bus or taxi. After climbing over a temporary wooden bridge, we found ourselves amid houses awaiting the decision to be destroyed or saved, depending on their state of deterioration. My informant said:

> This is like walking in a foreign country. I have never been here before. They call this the heart of Beirut, it is funny to see it like this. Look, there is a sign for Librarie Antoine! I didn't know they used to have a shop here. It is like a film, I don't feel this is real. I wonder what it was like here before the war. Which quarter are we in?

When he asked this, we were in Wadi Abou Jamil. Long before the war, it had been a Jewish neighbourhood. His only reference to it was in books he had read.

Public and private

Robert Park wrote, in 1916, that the newspaper is:

> the great medium of communication within the city, and it is on the basis of the information which it supplies that public opinion rests. The first function which a newspaper supplies is that which formerly was performed by the village gossip.
>
> (Park 1969 [1916]: 124)

The role attributed to the free press in today's discussions over civil society seems to find a precursor in these statements, as Park also notes that people's personal lives are only written about when they become interesting or important to the public. The city wedges in a forceful distinction between the private and the public that did not exist to the same extent in the village. If the village represented proximity and some civic intimacy, city life brought with it a distance between inhabitants and opened up a space between them in which the private would have to be hidden from view. Whatever we may today think of Park's sharp polarization between village and city, it does bring out an important idea with regard to space. If the city brings to the fore the differences that exist between its inhabitants and if the only uniting force lies in their roles as inhabitant or citizen, it becomes crucial to explore the spaces where they meet, encounter and confront each other.

I would like to stress, again, the fact that the war, which to a large extent created borders in the city of Beirut during the conflict, was not between neighbourhoods. It was between militias who had turned certain sections of the city into their strongholds. Thus the public spaces of these quarters had, to some extent, become more private in character. Certain streets would be blocked by their inhabitants to prevent unsolicited patrols or parking of booby-trapped cars. At the same time, the entrance of a home had become more permeable and open to neighbours and people in need. In short, the border between the public and the private changed.

Today, the shared spaces in the city centre, which used to serve as a sort of neutral meeting ground for people from all over the city, are going through a major transformation and reconstruction. The way it is done, the designs and plans proposed, have been subject to long and heated debates in Lebanon. Architects, intellectuals, politicians, NGOs, landowners and tenants have participated in discussions and debates that are far from over. One frequently heard criticism is that it seems odd that a private real-estate company should be given the mission to supervise the reconstruction of 'the heart of Beirut'.[8] Those in favour of this solution often reply to this question with another question: What was the alternative? This company, SOLIDERE, has done its best to show that it is acting in the interest of the public.[9] The people in possession of land within the area deemed 'SOLIDERE area' were given shares in the company in exchange for their rights to the plots of land on which the buildings stand.[10] Others were invited to invest in the company that provided SOLIDERE with the necessary capital. In addition to supervising all of the construction in the area, SOLIDERE is also constructing some buildings itself. In exchange for building roads and providing public space in the area, SOLIDERE has been given the rights to the land earmarked to become the new financial district. This is a landfill site that was a dumping site for rubbish during the war but is now being filled with the debris of the destroyed buildings.

The plans are at times grandiose and at times pragmatic. The head of the Urban Planning Department of SOLIDERE writes of the downtown area:

> This district, that used to be the common space for all Lebanese before the war, became a buffer space between them during the war. And since it was not hegemonized by any religious subgroup prior to the eruption of the war, it was naturally destined to become their battlefield.
>
> (Kabbani 1998: 245)

The word 'naturally' here underlines the problem and defines a straightforward history. Just as 'naturally' this area will become the new common space of Beirut and provide a neutral bridge for the various communities:

> The Central District ... will re-adopt its natural role as the city's meeting point. An outcome of central area reconstruction will be to re-establish the city centre as the natural arena for all of the country's diversified communities. Traditionally cosmopolitan, the central area remains the one unaffiliated location, where such mixing can naturally take place. Beirut's new Central District can, therefore, play an important role in the healing

process and in helping to rebuild a national and pluralistic identity. The city centre must provide a social arena, drawing people in and enabling all sections of the community to mix and meet.

(Gavin and Maluf 1996: 34)

Ideas for things to be provided in this public space range from pedestrian zones, heritage trails, parks and promenades to just empty lots. 'Markets will also be reintroduced into the residential areas in order to allow and encourage use by street vendors' (Kabbani 1998: 257).[11] The company inevitably plays a difficult role. It has to present itself as the guardian of public interest, the provider of spaces to promote peace and harmony, besides maintaining intercommunal communication, and at the same time be attractive to investors and the stock market.

What are the physical 'meeting points' in today's Beirut? The street is still a space for daily encounters and has also become, to an extent, a mode of social control *à la* Jane Jacobs,[12] thanks to the large number of people in the street and their apparent interest in what others are up to. As noted above, however, many of these streets are marked as belonging to or being under the control of a certain group. In certain quarters one can erase the writing on the walls to unearth the history of control of the area, as the slogans and symbols of one group are superimposed on those of another.[13] While these writings and symbols may make certain visitors feel uncomfortable, that is not always the case.

One of the most frequently mentioned functional public spaces is the Corniche, which runs along the coastline, west of the central district. It is now a few kilometres long, but was originally planned as a small promenade by the Ottomans in 1885. The French extended the Corniche to recreate the Promenade des Anglais in Nice (Ghorra-Gobin 1997: 100). The period after independence saw further additions to it. It took on its present form in 1974, after a strip was added which, in part, cut a fishing port off from the sea. This promenade is a space that opens up to the sea and it is a popular place to come for fishing, jogging or spend a few hours with friends. Except for a few houses on the inland side of the road that are occupied by Syrian soldiers, it is rather void of any group markings. A woman living in the east of downtown told me:

> It was one of the few areas that I really missed during the war. You know, it was on the other side of the centre and during the fighting it was impossible to go there. When there was a ceasefire it was the first place I went to. I have friends that live on the other side and we used to meet there. Now, after the war, I try to go there whenever I can, to walk and breathe some fresh air. I meet my friends and we walk for a while, for exercise. My sons had only heard about it and I told them they should go. There is not really any other place to run or walk. They were not scared but felt uncomfortable at first, but now they go more often than I do.

The Corniche is celebrated as *the* public space of Beirut, where people from all walks of life come to walk or sit for a while. It is considered by those in charge of planning as an example of civility reigning in a public space in Beirut. In the plans for the new downtown, the Corniche will be expanded down to the landfill site where it will become a sixty-metre-wide seaside promenade. This area is not yet constructed and it is difficult to predict whether it will be an extension of the Corniche or something rather different. One important aspect of the Corniche is that its whole width is visible at all times; one can see everyone passing by, so that in this respect it resembles a large sidewalk. This effect will surely be missed when its width is increased, but then again the increase in size may very well allow other kinds of activities for which the Corniche, as it is, would be too narrow.

One of the districts that borders the Corniche is Ayn el-Mreisse, which is also a boundary of the downtown zone and quite a poor neighbourhood with large sections of badly dilapidated buildings, especially in the quarters next to the Hotel District. In large portions, Ayn el-Mreisse has been cut off from the sea through the gradual expansion of the Corniche. Commenting on the Corniche, one man made it clear that he did not go there: he stayed on his side of the road, in his quarter. He considered the Corniche to be for people from outside, meaning people from other parts of the city. Unless he was going fishing, he would stay on his side of the road.

Public space: neutral or political?

In November 1998 a large concert was organized in what used to be the heart of the city centre, known as 'the Martyr's Square' or the Bourj. It was an event that brought many famous and popular singers together to sing a song called 'The Arab Dream'. The song was accompanied by a twenty-minute video showing images from all of the wars that had been fought on Arab territory since 1948. It was an odd song in the sense that it was a far cry from the pan-Arabism of Nasser and yet reminiscent of the recent painful history that united most Arab countries, creating sentimentality for the unification of the Arabs in spirit. It is noteworthy that the video did not show a single image from the civil wars in Lebanon.

SOLIDERE had encouraged the concert be held in 'their' area to show that it was indeed the 'natural' meeting point for all Lebanese. A few hundred thousand people turned up for the event.[14] There was a strange atmosphere. For the first time since the end of the war, people were tolerant or felt free to walk around carrying signs with political slogans or pictures of their leader. Demonstrations are still not allowed in Lebanon and at this event people took the chance to show their allegiances under the pretext of 'the Arab Dream'. I was there with a young journalist who was quite shocked to see small groups, mostly of young men, walking through the crowd, waving their banners, since he had never before seen this on the streets of Beirut. The event ended in some minor chaos, as some people in front of the stage clashed with the police. The clashes had no political undertones, being mainly caused by the long waiting, growing impatience and a lack of seats.

At present, there are few places in Beirut that function as places for encounters. Apart from places of leisure, such as cinemas and shopping centres, and perhaps the Corniche, there are not many places where the 'other' can be seen without being identified as the 'other'. As the mixing and meeting of inhabitants seems to be high on the rhetorical agenda of many people in positions of power, one can perhaps hope that this will change. There is, however, a need for caution. Should it be neutral or political spaces that are created to serve as a meeting ground (Arendt 1963)? Unmarked neutral places are almost impossible to create. But it may be equally hard to provide public spaces that can assume a political character, especially in the context of Lebanon where politics and community identity have gone hand in hand. To create 'the political space [that] comes into being when people live together for the purpose of practising freedom' (Dossa 1989: 87) should perhaps be the goal of those wanting to bridge or eradicate the borders between communities. Hannah Arendt sees this 'space of appearance' as essential for the constitution of political beings who can become citizens actively engaged in society. However, it is not always necessary, or enough, to plan and create a public space in which citizens can appear in their political significance: 'Wherever people gather it is potentially there, but only potentially, not necessarily and not forever' (Arendt 1958: 199).[15]

One problem, which I have not brought up yet, is the aspect of class in a divided city. The city is for whom? The inhabitants of a city have a right to the spaces of that city. That is to say, the inhabitants of a city have a right to be present on the streets and to participate in the issues that concern that space to which they, in part, contribute and give functions. It also means a right to refuse to 'allow oneself to be removed from urban reality by discriminatory and segregative organization ... [and e]qually it stipulates the right to meetings and gathering' (Lefebvre 1996: 195).[16]

Class differences are nothing new in Lebanon, or anywhere else for that matter. But the urban diversity of Beirut may well take on even more sharp-edged class divisions in the near future. One of the most frequently heard criticisms against the plans and actions of SOLIDERE is that they are creating a downtown for the wealthy, a 'city' apart from the rest. Rents are expected to be high both for businesses and dwellings. Apartments are generally quite large in Beirut. In a recently renovated area downtown called Saifi Village, the apartments that are offered range from 146 square metres to 345 square metres and all residents will have access to a private fitness club including a beauty centre and a swimming pool. These apartments are being sold at US$1,750 per square metre – that is, the cheapest apartment will cost a quarter of a million US dollars. The rent for the same apartment would be US$16,060 a year.[17] In 1996 the average yearly income of a worker was US$4,164. This is well in line with one of the slogans of the company: 'SOLIDERE: Developing the *finest* city center in the Middle East'.

If we add class to the denominational configuration of the city, we may witness the development of a city centre which is indeed a 'mixed' space. A mixture of the upper strata of each denomination residing in the 'natural' meeting place of all Lebanese, surrounded by enclaves of more homogenous quarters. I believe that what should be developed, in looking at the city of Beirut, is a 'prism of citizenship'. By paying attention to the 'spaces of insurgent citizenship' (Holston 1996; see also Painter and Philo 1995), urban planners may be able to respond to the needs of the inhabitants of the city and not only to the need to attract investors to the city.

> The new spaces of citizenship which result are especially the product of the compaction and reterritorialization in cities of so many new residents with histories, cultures and demands which disrupt the normative and assumed categories for social life.
>
> (Holston 1996: 57)

This is not to say that a city can disregard international and regional financial development. However, to do so at the expense of the needs of the inhabitants is a different matter. Indeed, this is one of the most frequently heard criticisms of the projects in downtown Beirut: 'The downtown will not be for the Beirutis', a merchant who used to own a shop downtown told me. 'It will probably be for the rich people, mostly from the east.' At the moment, it is perhaps too early to judge.

The ephemeral character of public space is very evident in present-day Beirut, where the idea of public space has become a rhetorical vehicle for a profit-making company. The spaces for encounters do exist. Leaving the grand plans and their critics aside for a moment, it seems certain that civility (Sennett 1992) will find space to continue to prosper or refuse to be erased. The public space is not always an arena for reducing the number of identities of a person, nor is it necessarily a place where private aspects of a person go uncommunicated. The idea of citizenship is brought up here to point to the possibility for learning, in public, that one has interests in common with others. Identity is not something that one brings from the private realm into the public but, rather, is formed by participation in public life (Calhoun 1993). The diversity that seems intrinsic to any definition of the city lies at the heart of the possibility of creating actively concerned citizens. On the wall next to a secretary working in the urban planning department of SOLIDERE, there was a piece of paper with a sentence on it that to me, ironically, commented on the new borders being created in Beirut. It was a famous quotation from Aristotle: 'A city is composed of different kinds of men; similar people cannot bring a city into existence.'[18]

Notes

1 In connection with Expo 2000 in Hannover, there was recently a rather large and pompous conference in Berlin, 'Urban 21', which will surely result in a glossy publication attesting to the importance of the issues discussed.

2 There is a good example of the liberating and inhibiting effect of the 'city air' in the film *La Vita é Bella* by Benigni. Two men arrive in the city and one takes a deep breath and says to the other that here in the city one is free to do as one wants. The second man then cries out in joy and leaps up and down. The first man turns to him and snaps, 'What are you doing? Behave! We are in the city now.'
3 By 'peri-urban' Saliba is referring to a kind of urban fabric between the urban and the suburban, with clearly defined streets, small/medium lots set back from the street and medium surface exploitation.
4 See al-Shaykh's *Beirut Blues* (1995) or the film *West Beyrouth* by Ziad Doueiri.
5 This area does not appear on the map but includes the areas of Remeil and Mar Nqoula. This approximation is in keeping with popular usage and the area of activity of the Comité de developpement de Gemmayze.
6 He had made his fortune by choosing to expand instead of abandoning his printing press during the early years of the fighting in the city. Thanks to his contacts within a Christian militia, he was warned before any bombardment or fighting so that he could send off his products and close his facilities.
7 The name is a French acronym for 'The Lebanese Company for the Development and Reconstruction of Beirut Central District'.
8 The creation of SOLIDERE was endorsed by the government in 1992 and its masterplan for the assigned area approved by the Council of Ministers the same year. The company is considered to be the brainchild of Rafiq Hariri (Prime Minister 1992–6, 1996–8; he resigned in December 1998, but was re-elected in autumn 2000). Hariri also owns 10 per cent of the shares in the company, which is the maximum allowed for any one person.
9 A glance at the members of the board of the company reveals that all major denominations are represented.
10 Another point of criticism is that the shares they were given are considered to be well below the market value of their lots of land.
11 No visitor to Beirut could fail to notice that the street vendors have no problem serving the various neighbourhoods, even those lacking a marketplace. They move around the city with their carts, announcing their produce as they move.
12 Jane Jacobs has argued that a successful city must make a person feel safe on the street when surrounded by strangers. This is, in her view, achieved though three qualities:

> First, there must be a clear demarcation between what is public space and what is private space ... . Second, there must be eyes upon the street, eyes belonging to those we might call the natural proprietors of the street ... . And third, the sidewalk must have users on it fairly continuously, both to add to the number of effective eyes on the street and to induce the people in buildings along the street to watch the sidewalks in sufficient numbers.
>
> (Jacobs 1996: 106)

13 The urban planner mentioned above, working for SOLIDERE, once stated that there is not really that much to renovating a building or neighbourhood: 'All you need is a bucket of paint.'
14 Again, estimates of the number of people vary.
15 Benhabib has drawn attention to one interpretation of Arendt, pointing out that public space in her terms is:

> that space in which only a certain type of activity ... takes place, or is delimited from other social spheres by reference to the substantive content of the public dialogue.
>
> (Benhabib 1992: 81)

16 The original text is from *Le droit à la ville* (Paris: Anthropos, 1973).
17 Prices are from the internet homepage of SOLIDERE (http://www.solidere.com.lb) where one can also see the layouts of the apartments and their location in the city.
18 The quote was without reference on the office wall but it had been taken from Aristotle's *Politics*. In the version I have, the full quote is:

> The state [*polis*] consists not merely of a plurality of men, but of different *kinds* of men; you cannot make a state out of men who are all alike.
> (Aristotle 1992: 104)

References

al-Shaykh, Hanan (1995) *Beirut Blues*, London: Chatto & Windus.
Arendt, Hannah (1958) *The Human Condition*, Chicago: University of Chicago Press.
—— (1963) *On Revolution*, London: Penguin Books.
Aristotle (1992) *Politics*, trans. T.A. Sinclair (1962), revised Trevor J. Saunders (1982), London: Penguin Books.
Benhabib, Seyla (1992) 'Models of public space: Hannah Arendt, the liberal tradition, and Jürgen Habermas', in Craig Calhoun (ed.), *Habermas and the Public Sphere*, Cambridge, MA, and London: MIT Press.
Calhoun, Craig (1993) 'Civil society and the public sphere', *Public Culture* 5: 267–80.
Danger *frères* (1932) *Ville de Beyrouth: Rapport d'enquête et justificatif*, Paris: La société des plans régulateurs de villes.
Dossa, Shiraz (1989) *The Public Realm and the Public Self: The political theory of Hannah Arendt*, Waterloo: Wilfrid Laurier University Press.
Figuié, Gérard (1997) *Le point sur le Liban 1998*, Beirut: Anthologie.
Gavin, Angus and Maluf, Ramez (1996) *Beirut Reborn: The restoration and development of the Central District*, London: Academy Editions.
Ghorra-Gobin, Cynthia (1997) 'Beyrouth ou les conditions d'émergence de l'espace public', in Michael F. Davie (ed.), *Beyrouth: Regards croisés*, Tours: URBAMA.
Holston, James (1996) 'Spaces of insurgent citizenship', *Architectural Design* 66(11–12): 54–9.
Jacobs, Jane (1996) 'The uses of sidewalks: safety' (repr. from *The Death and Life of Great American Cities*, 1961), in Richard T. LeGates and Frederic Stout (eds), *The City Reader*, London: Routledge.
Kabbani, Oussama (1998) 'Public space as infrastructure: the case of post-war reconstruction of Beirut', in Peter Rowe and Hashim Sarkis (eds), *Projecting Beirut: Episodes in the construction and reconstruction of a modern city*, New York: Prestel-Verlag.
Khalaf, Samir (1993) 'Urban design and the recovery of Beirut', in Samir Khalaf and Philip S. Khoury (eds), *Recovering Beirut: Urban design and post-war reconstruction*, Leiden: E.J. Brill.
Lefebvre, Henri (1996) *Writings on Cities*, selected, trans. and intro. E. Kofman and E. Lebas, Oxford: Blackwell.
Low, Setha (1996) 'The anthropology of cities: imagining and theorizing the city', *Annual Review of Anthropology* 25: 383–409.
Makdisi, Ussama (1996) 'Reconstructing the nation-state: the modernity of sectarianism in Lebanon', *Middle East Report* July–September: 23–6.
Mouzoune, Abdelkrim (1999) *Les transformations du paysage spatio-communautaire de Beyrouth (1975–1996)*, Paris: Publisud.

Nasr, Salim (1993) 'New social realities and post-war Lebanon: issues for reconstruction', in Samir Khalaf and Philip S. Khoury (eds), *Recovering Beirut: Urban design and post-war reconstruction*, Leiden: E.J. Brill.

Painter, Joe and Philo, Chris (1995) 'Spaces of citizenship: an introduction', *Political Geography* 14(2): 107–20.

Park, Robert E. (1969) 'The city: suggestions for the investigation of human behavior in the urban environment' (repr. from *The American Journal of Sociology* 20 (1916)), in Richard Sennett (ed.), *Classic Essays on the Culture of Cities*, New York: Appleton Century Crofts.

Saliba, Robert (1998) *Beirut 1920–1940: Domestic architecture between tradition and modernity*, Beirut: The Order of Engineers and Architects.

Salibi, Kamal (1988) *A House of Many Mansions*, London: Tauris.

Sennett, Richard (1992, 1st edn 1974) *The Fall of Public Man*, London: W.W. Norton.

—— (1994) *Flesh and Stone: The body and the city in Western civilization*, London: W.W. Norton.

Wirth, Louis (1938) 'Urbanism as a way of life', *The American Journal of Sociology* 44: 1–24.

Part II
The meso-level analysis of urban ethnic encounters
The neighbourhoods of the cities

6 Perception and use of space by ethnic Chinese in Jakarta

Harald Leisch

Economic power of an ethnic minority

Why is the Chinese population of special interest in Indonesia and especially in Jakarta? Although about 20 per cent of all overseas Chinese live in Indonesia, they are only about 3 per cent of the country's population and in Jakarta maybe about 10 per cent (Poston, Mao and Yu 1994). They are an ethnic minority, but not an ordinary one.

Today, most of the Chinese have Indonesian nationality and many have mixed with the *pribumi* (aboriginal) Indonesian population through marriage and by adopting their culture. Such people are called *peranakan*, a term used in referring to 'the Chinese who are either children of mixed marriages or have been born and raised in Indonesia'; they have never even been to China nor have they been educated in the Chinese culture. The economic activities of the *peranakan* Chinese are of minor importance, since most of them seek white-collar jobs in companies and are not really interested in running their own businesses.

However, the Indonesian economy is mainly in Chinese hands. Some estimate that the Chinese share in the corporate wealth of Indonesia was around 70 per cent of the total, before the economic crisis of 1997 and the subsequent eventful downfall of President Suharto in 1998 (Mackie 1999: 187). This might be an overestimate, but it explains the economic importance of the rich Chinese group who are, in fact, only a small fraction of the Chinese in the country, and are mostly excluded from any kind of governmental occupation and from political power in Indonesia. The majority of the Chinese who control important conglomerates have a closer relationship with their motherland. Of these, the group called *totok* (Chinese with Chinese parents or those who have been born and raised in China) often receive Chinese education and still use their native language in their business. Because of the restrictions on the economic activities of non-native Indonesians in the 1950s and the 1960s, the *totok* Chinese have formed what are known as 'Ali-Baba' alliances with *pribumi* entrepreneurs ('Ali' standing for the *pribumi* and 'Baba' for the *totok* partner). The Chinese economic elite had to find ways to collaborate with the Indonesian power elite in order to get licences or permits for their businesses. The Chinese who found friends in high political positions during the New Order Period are called *cukong*, which means 'master' in the Hokkien dialect. Most of the *cukong* businessmen are of *totok* background and a large number of them do business in banking, finance, real estate, import–export, consumer goods, manufacturing, timber, construction, transportation and even the typical Indonesian clove-cigarette industry (Suryadinata 1997).

A rich minority, in contrast to a comparatively poor majority, is often a basis for jealousy. As will be seen later, the Chinese have frequently been attacked by indigenous Indonesians who tried to get rid of them, although the latter group is economically partly dependent on Chinese business. But when people are hungry and cannot pay for their daily rice, as was the case during the economic crisis in 1997–8, rage is incited against the small rich part of the population. When *agents provocateurs* take advantage of such a situation, people act without thinking. The attacks on Chinese property and people show that it is not without reason that the Chinese people in Indonesia are scared, especially when they are outside their own community. This fact, which will later be described for the Chinese in Jakarta in more depth, is important for the understanding of Chinese behaviour in their perception of urban space. Before this is analysed, it is necessary to look back to the origins of the Chinese in Indonesia.

Chinese overseas migration and settlement in Indonesia

International migration is often connected to international trade, which indeed was the main reason for major Chinese migration to overseas countries during the last two centuries. Wang Gungwu (1991) distinguishes four major patterns of Chinese migration:

Huashang is the earliest and still dominant type of migration. Chinese merchants and artisans settled down in the most important trading points, mainly coastal towns, and set up businesses there. Until 1850 *huashang* was the main pattern of emigration to other Asian countries, particularly to Southeast Asia, and it remains the basic type of Chinese overseas migration.

Huagong is the migration of Chinese coolies. *Huagong* existed from the late eighteenth century, becoming important in East Sumatra shortly after 1819, and was the dominant pattern of overseas migration between the 1850s and 1920s. North America and Australia, where coolies were engaged in gold mining and railway building, were the preferred destinations. But mines in Kalimantan (Indonesia) also attracted coolies, who often returned to China after their contracts ended.

Huaqiao ('Chinese sojourner') describes the migration, mainly dominant in Southeast Asia until the 1950s, of well-educated professionals who left China to educate the children of Chinese immigrants.

Huayi ('Chinese descent') is the migration of Chinese – already living and born overseas – to a further foreign country. It has been prevalent since the 1950s, especially after the indigenous populations of some Southeast Asian countries indicated, more or less clearly, to the Chinese that they were no longer welcome.

Huashang was the pattern of Chinese immigration in Indonesia for hundreds of years. By the time Europeans arrived in the area in the sixteenth century, Chinese quarters (*pacinan*) had already been well established in some towns. The 1920s in Indonesia, with its strong economic growth, attracted many migrants from China – in particular, the coolies. It was the last big wave of immigration. While the merchants mainly settled down in the cities in Java, coolies mainly worked in the outer islands. With a distinction between Europeans, Foreign Orientals and indigenous people, colonial apartheid had legally been institutionalized by that time. According to the colonial census of 1930, 1,233,000 Chinese lived in the Dutch Indies; nearly half of them in Java (Heidhues 1998). After the violence against the Chinese in 1942 and 1945, the government had the feeling that there were too many Chinese in the country and therefore did not admit Chinese immigrants from the 1950s onwards. As a result of governmental measures to limit the economic activities of the Chinese and other foreigners, some Chinese emigrated to China while a considerable number remained in Indonesia.

During the anti-Communist riots from 1965 to 1967, when Suharto took over the presidency, several thousand Chinese lost their lives, though they had not been the main target of the violence. Again, several hundreds of thousands of Chinese left Indonesia. During the Suharto period, all Chinese schools were closed, Chinese newspapers – even Chinese characters on houses – were forbidden and the Chinese were asked to change their names to Indonesian-sounding ones. Thus, during the last thirty years, the Chinese people have not been able to preserve their cultural identity – at least, not openly. The assimilation efforts have been in part successful. Most of the younger Chinese have adopted the Indonesian culture. In fact, many of them no longer know much about their original culture outside those who have had private tuition from Chinese teachers or have been sent to China for their education. However, most Chinese parents, if they can afford it, send their children to universities abroad or to private universities in Indonesia.

Astonishingly, the Chinese culture has survived. This can, for instance, be seen in the traditional lion parade, the Barongsai, which is performed to celebrate the Chinese New Year or on other occasions. The Barongsai could not be staged on the street during the Suharto presidency, but under the government of President Abdurrahman Wahid (October 1999–July 2001) anti-Chinese regulations were loosened. There is no longer a special code in the passports of Chinese people – of whom an estimated 90 per cent are already Indonesian nationals – indicating their identity and the Chinese are allowed to preserve their culture again. Even the first Chinese daily, the *Zhinan Post*, was back in circulation in early 2000. In the meantime, about half a dozen Chinese newspapers and magazines had been published in Indonesia, creating competition among themselves (*The Jakarta Post* 30 April 2000).

The riots in 1998 resulted in a huge emigration of Chinese people to Singapore, Australia and Canada, or to more secure areas within Indonesia, such as the mainly Protestant region of North Sulawesi or the tolerant Hindu area of Bali. In spite of the memory of this horrible violence against the Chinese population, most of the Indonesian Chinese returned after Wahid's election as president in October 1999. They brought back their money, too. The chairman of the Indonesian Business Development Council, Sofyan Wanandi (a Chinese), estimates that 25 per cent of the total domestic funds that had been deposited overseas during the political–economic crisis of 1997–8 (around US$15 billion) has returned to Indonesia (*The Jakarta Post* 23 February 2000). Once again, a lesson should have been learnt from history that the economic activities of the Chinese are highly important for the development of the country and, in particular, the capital Jakarta.

Development of Chinese communities in Old Jakarta

The Dutch colonial rulers realized that the small harbour town Jayakarta would have a dominant position in the Indies trade, due to it being strategically located between the Indian Ocean and the South China Sea. In 1619 the Dutch changed the name of the town to Batavia, which was renamed again as Jakarta in 1945, during the struggle for independence. Jan Pieterszoon Coen of the Dutch East India Company (VOC) planned the further development of the town. He was well aware of the fact that the development of the town would be dependent on Chinese economic activities (Abeyasekere 1989).

The European residents in Batavia were steadily outnumbered by ceaseless Chinese immigration brought on by economic development. In 1740 growing panic on the Dutch side resulted in a massacre, during which most of the Chinese lost their property and many paid with their lives. After these riots, the Dutch no longer wanted the Chinese to live in their immediate vicinity, within the walled area. Thus they created a new quarter for them outside the town wall, called 'Glodok', which is still at the heart of Jakarta's Chinatown (Abeyasekere 1989). Nevertheless, the Dutch soon realized that they needed the Chinese; as a consequence, Chinese immigration started again. The Chinese remained an important factor in the business deals of the Dutch. The Chinese had to pay more taxes than the *pribumi* residents, which was one of the reasons for the colonial rulers to keep the Chinese separate from the local population. 'Repeatedly, colonial regulations emphasized their separateness; they had to live in separate quarters, dress in a Chinese way ... ' (Heidhues 1998: 157). The Dutch also introduced a 'pass-and-quarter system', which allowed the Chinese to reside only in the Chinese quarters and excluded them from travelling freely to other areas outside their own town quarters. Later, these regulations were relaxed.

In 1959 – that is, about a decade after independence – a new regulation prohibited the Chinese (as well as other foreigners) from retail trade outside the major towns. This forced them to live in the cities, where their concentration has remained the highest, with about 30 per cent in the city of Pontianak (Kalimantan), more than 10 per cent in Medan (Sumatra) and an estimated 10 per cent in Jakarta (Heidhues 1998).

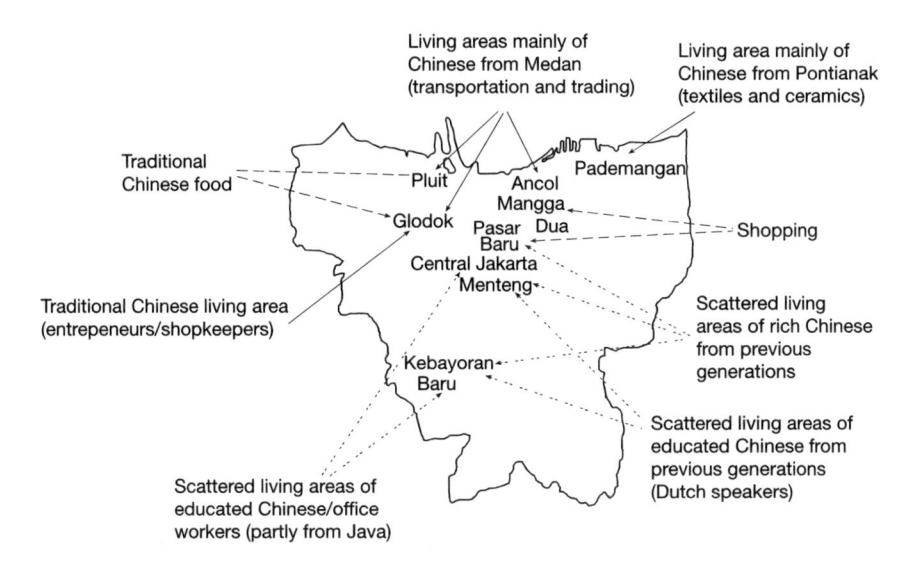

Map 6.1 Chinese space in Jakarta
Source: H. Leisch, after interviews with Chinese residents

As a consequence of the developments mentioned above, there has always been a clearly defined space for the Chinese in Old Jakarta, a neighbourhood adjacent to the old core of Batavia which has persisted until today (Map 6.1).

'Chinese' space in Jakarta – persistence and change

Although the Chinese quarters have always been well known in Jakarta, the Suharto government tried to 'conceal' the neigbourhood by banning the use of Chinese characters. Thus, during his term, Glodok (Jakarta's main Chinatown in the Kota district) could not be recognized as such by visitors to the city.

Easily recognizable from the outside or not, the Chinese quarters have persisted, stretching out to the west and north and partly to the south of Jakarta. Today the quarters Glodok, Pluit, Ancol, Pademangan, Mangga Dua and Pasar Baru are the main 'Chinese' areas within the boundaries of DKI Jakarta (see Map 6.1). New gated quarters, like Puri Indah, have been developed in the west of Jakarta, where some people estimate that the percentage of the Chinese make up about 80 to 90 per cent of the population.

The different Chinese quarters are not culturally homogenous, showing variation in the origins, occupations and attitudes of their inhabitants, not to mention in the different daily activities that take place in specific areas. This is typical not only of the Chinese quarters in Jakarta, but also for cities in China. Gaubatz (1998) has shown that in Guangzhou in China, a spatial separation exists within the town based on the occupations and origins of its residents. In most cases there is a correlation between occupation and origin. This also applies to Jakarta (see Map 6.1). For example, the Chinese coming from Medan

(Sumatra) usually live in the areas of Glodok, Pluit and Ancol, and mostly set up business in transportation and trading. Pademangan is the main area for the Chinese coming from Pontianak (Kalimantan) who work in the textile and ceramic industries.

Glodok still is the most traditional Chinese quarter in Jakarta. It is strewn with restaurants and small shops that provide traditional Chinese food and other Chinese goods. The bigger shops specialize in electronic goods. Glodok is dominated by *totok* businessmen, the most traditional of the Chinese, who are entrepreneurs or traders. The majority speaks Chinese languages, not being familiar with the Indonesian language. This is also why they concentrate their daily activities within their own quarter. It is also for security reasons that the Chinese people prefer to live in segregated quarters where they can protect themselves with fences and barbed wire. But the spatial concentration of Chinese is also a danger, because they make an easy target to attack. This became clear during the riots in May 1998, when shops (and also some domiciles) in Glodok were looted, buildings destroyed and people attacked. The reconstruction of the Glodok market started in the year 2000, at a cost of around US$7 million (*The Jakarta Post* 21 February 2000).

The older people remain in Glodok, whereas their *peranakan* children, who have partly adopted the Indonesian culture, move to other areas and now live scattered in central and south Jakarta, mixed with the indigenous Indonesians. Neither these *peranakan* nor their parents in Glodok belong to the really rich group of Chinese in Jakarta.

The rich Chinese, whose wealth had been accumulated by previous generations, live in elite quarters that were formerly the residential areas of the Dutch during the colonial period, like Menteng and, later, Kebayoran Baru (Abeyasekere 1989). The Chinese living in these neighbourhoods are usually well educated and speak Dutch, a status symbol during the colonial period and even afterwards. Menteng has always been a residential area for the upper class. Shortly after the transfer of sovereignty by the Dutch to the Indonesians in 1949, the economically better-off Chinese moved to this area to live, mingling with the *pribumi* and European elite already there. At present, these Chinese continue to live together with indigenous Indonesians who work in top government positions, using the spacious neighbourhood to make the business connections necessary in dealing with the government. Not surprisingly, a great number of this group are known as *cukong* businessmen – that is, those with friends among the Indonesian power elite.

The Chinese not only prefer to have their own quarters for living, but they also as far as possible choose Chinese-dominated areas for their daily activities. Many Chinese make long trips through Jakarta, just to have traditional Chinese food in Glodok. When they go shopping, they tend to choose malls in 'Chinese areas' like Mangga Dua and Pasar Baru. Shopping malls like Taman Anggrek, Citraland or Puri Mall in the west and Mega Mall in the north of Jakarta are most suitable for the Chinese, because these shopping centres are Chinese dominated. In contrast, Blok M in the south or Atrium in central Jakarta have mainly *pribumi* customers and are, therefore, often not fancied by the Chinese. However, this does not mean that Chinese people avoid all neighbourhoods outside the Chinese areas. In fact, they only steer clear of those quarters (like the harbour, Tanjung Priok or Senen) that are also avoided by the indigenous Indonesians because they are known as *rawan* (insecure) and people are scared of muggers.

Most Chinese, however, feel that the east of Jakarta, where the majority of the population is Muslim, is also not a safe zone for them. History has taught them that they can never feel secure outside their own community. Most of them state that they have been scared since childhood and have always been warned by their parents to avoid activities outside the Chinese community. Despite that, it is not correct to speak about a 'community', in the true sense of the term, in some Chinese residential areas: they often do not know their neighbours at all. They only live in the same neighbourhood because they feel safer that way.

Chinese land development in Jabotabek

Since security has become such an important factor in Chinese spatial behaviour, it must consequently be the basis for any new land development. Chinese developers own huge areas of land outside Jakarta, especially in the city of Tangerang, which forms one mega-urban agglomeration with Jakarta, usually referred to as Jabotabek. This situation seems almost like a repetition of history, because in the late eighteenth and early nineteenth century, the government in Batavia had tried to gain revenue by selling land. The main buyers, besides Europeans, had been Chinese. They not only bought land within Batavia but also huge areas in Tangerang (Abeyasekere 1989) where they cultivated the land for agricultural production. This group of Chinese is still known as *Cina Benteng* – a closed society, more or less separate from the Chinese in Jakarta.

Today, Chinese conglomerates buy land, though this time mainly from the villagers and for a very cheap price. They develop new towns, which are, of course, not only for the Chinese. But since these new towns also provide high security standards in 'totally protected zones', the economically better-off Chinese prefer to live there. Usually, the villagers have no idea about the enormous increase in value of their land until after the development projects are complete. When, occasionally, the villagers realize how cheap they have sold their land, they get angry and try to attack the staff of the company carrying out the construction. With a holding of around 6,000 hectares, Bumi Serpong Damai is the biggest town, although until now only around 1,300 hectares have been developed. With 2,800 hectares, the most exclusive new town seems to be Lippo Karawaci, which is being developed by a company belonging to the Lippo Group.

Both new towns are examples of a new kind of land development in the agglomeration of housing for Jakarta, because they provide a ready-made infrastructure with schools, shopping facilities, sports clubs, cinemas and a hospital.

Only middle- and upper-class people can afford to buy houses there. Similarities to the development in China are obvious, with only upper-class people able to buy houses in gated communities in the suburbs (Gaubatz 1998). In both types of agglomeration of housing, the style of the new towns is strongly influenced by American architecture and even by the American way of life – albeit modified in an Indonesian or Chinese style, with references to *feng shui*.

As land developers of such new towns, the Chinese create their own urban space, which, in turn, makes them targets for jealousy and violence. During the riots in 1998, the Lippo Super Mall in Lippo Karawaci – at the time also owned by the Lippo Group – was looted and partly destroyed, causing US$25 million damage. Since it re-opened in September 1999, it has no longer been a shopping mall for the upper and upper-middle classes, but for the lower classes, mainly indigenous Indonesians; its principal shares were sold to a Dutch company. During the riots in 1998 and for almost a year afterwards, an army tank protected the main entrance of the residential area. But there is always danger: the people from the surrounding villages hardly profit from such new towns. Most of the employees in the Super Mall and the maids in the residential area come from other parts of Jakarta, or elsewhere in Indonesia, but not necessarily from the surrounding villages. The villagers are not allowed to sell any goods as hawkers, at least not in Lippo Karawaci – there they are prohibited from entering the residential area. However, the villagers are occasionally confronted by the world of luxury of the new town – for example, when they try to join in the entertainments in the Mall on Sundays. In theory, various forms of entertainment are open to the villagers; in practice, they cannot afford them.

Although the Chinese residents do not have any contact at all with the surrounding villages, they are aware of the jealousy among the villagers. Therefore, the company that is in charge of Lippo Karawaci, as well as the one that runs Bumi Serpong Damai, donates gifts (such as free chicken) to the villagers. The company also permits the villagers to cultivate land that is not yet developed for housing. The companies prefer to disturb the villages as little as possible, such as avoiding diverting water from the surrounding irrigated rice-fields. Nevertheless, not all conflicts can be avoided. For instance, although every neighbourhood is protected by walls and fences, besides being guarded by security posts at each gate and inside, cases of theft still occur. Often the thieves are among the maintenance workers. For that reason, the land developers now choose to send workers home each night.

Conclusion

Since the Chinese began to migrate to Jakarta, they have adopted a special role in its development. The economic prosperity and overall development of the city have been more or less dependent on the Chinese population and their economic activities. In Jakarta, the Chinese population has always lived segregated, partly by choice and partly because the administration forced them to. As foreigners, they preferred to live within their own community, which gave them the feeling of being at home, warm and secure. The government, on the other hand, tried to control them by confining them to certain areas in the city. During the colonial period, the Chinese and the *pribumi* middle class lived in urban neighbourhoods that still had a rural character, in houses that were not suitable for them (Wertheim 1959). But with increasing access to education, a Chinese bourgeoisie came into existence and the wealthier Chinese were able to leave the Chinese neighbourhoods and live together with the European bourgeoisie. To the present day, the Chinese and *pribumi* Indonesians mix only within the (upper-)middle and upper class, and only as long as they have higher education and/or wealth in common.

The Chinese in Jakarta still do not feel safe outside their communities and prefer to live in their quarters, undisturbed by others. This situation implies a high security risk. An ethnic group that lives concentrated in one spot can easily be singled out and become subject to violence, which has happened more than once. Spatial concentration is also a dilemma for the government, because it allows control over the Chinese but, at the same time, undermines plans to force them to assimilate with indigenous Indonesian society so that the Chinese become 'invisible' as a category.

In this respect, a considerable number of the Chinese have mixed with the *pribumi* people, though this may be limited with regard to class and space. The public space in Jakarta is not really public but clearly divided into public space for different groups of society, like public space for the *pribumi* upper and upper-middle class, public space for the *pribumi* lower class, public space for the (upper- and middle-class) Chinese, and so on. It is evident that people with different status and/or different ethnic origin do not want to mix more than is necessary and they usually respect the 'privacy' of the space of other groups. Nevertheless, space with an exclusively Chinese population still exists. Within their own communities, the Chinese can and do provide their own security, thus declaring that part of the public space is, more or less, private. Whether or not these neighbourhoods are surrounded by walls, fences and barbed wire, their privacy is normally respected by other groups of society.

On the other hand, such spatial exclusion implies the concentration of power and wealth in a developing country. Hence, jealousy among the poor people living around such protected 'islands of wealth' cannot be avoided. It is only a question of time before jealousy turns into crime, clashes and violence. The Chinese who live in their own protected spaces are very much aware that they live in a golden cage. To help themselves feel better, they try to make their cage more comfortable by locating an increasing variety and number of facilities inside their settlements, making it unnecessary to leave their world. The exclusive new towns that are being developed by and mainly for the Chinese are the best examples. As long as the population in general cannot profit from economic growth, the Chinese enclaves in Jakarta will have to be protected by high walls and peace offerings – in kind, money or through the temporary use of land – to the impoverished neighbouring villagers.

Acknowledgements

The fieldwork for this chapter would not have been successful without the help of many Chinese residents of Jakarta. The author is extremely grateful to all of them.

The author would like to thank the Deutsche Forschungsgemeinschaft (DFG) for their financial support, which made his research in Indonesia (project no. LE 1010/1) possible.

References

Abeyasekere, Susan (1989) *Jakarta: A history*, Singapore, Oxford and New York: Oxford University Press.
Gaubatz, Piper (1998) 'Understanding Chinese urban form: concepts for interpreting continuity and change', *Built Environment* 24(4): 251–70.
Heidhues, Mary Somers (1998) 'Indonesia', in Lynn Pan (ed.), *The Encyclopedia of the Chinese Overseas*, Singapore: Archipelago Press.
Mackie, Jamie (1999) 'Tackling "the Chinese problem"', in Geoff Forrester (ed.), *Post-Soeharto Indonesia: Renewal or chaos?*, Singapore: Institute of Southeast Asian Studies.
Poston, Dudley L., Jr, Mao, Michael Xinxiang and Yu, Mei-Yu (1994) 'The global distribution of the overseas Chinese around 1990', *Population and Development Review* 20: 631–45.
Suryadinata, Leo (1997) *The Culture of the Chinese Minority in Indonesia*, Singapore and Kuala Lumpur: Times Books International.
Wang Gungwu (1991) *China and the Chinese Overseas*, Singapore: Times Academic Press.
Wertheim, Willem Frederik (1959) *Indonesian Society in Transition: A study of social change*, Westport: Hyperion Press.

7 Urban fear in Brazil
From the *favelas* to *The Truman Show*

*Carmen Sílvia de Moraes Rial
and Miriam Pillar Grossi*

The theme of this chapter is the growing fear of urban violence, the fear experienced by large numbers of residents of different social class and ethnic backgrounds in Brazil's large and medium-sized cities. We deal here primarily with the strategies of protection from criminals, as developed by urban residents, and how these have brought about profound changes in the spatial and ethnic configuration of the city, as well as in attitudes, conduct and social practices in public space. These attitudes have given rise to a type of urban culture in Brazil that has been recognized in the social sciences as a 'culture of fear'.

In our understanding and according to the Brazilian researchers who coined the term, 'culture of violence' refers to the complex of social and symbolic practices adopted by the residents of Brazilian urban centres in order to confront violence. In their usage of the term 'violence' (an indigenous categorization), the residents of the city conflate different types of social action carried out in diverse contexts, from murders committed during muggings to aggression against women and children.

The data for this article were obtained primarily from the systematic reading of Brazilian newspapers (as referred to in the notes) over the past five years[1] and interviews carried out with urban residents, as well as the direct observation of life in several major Brazilian cities. These cities are Rio de Janeiro and São Paulo (which both have more than five million inhabitants and rank highest in the statistics on violence in the country), Brasília and Florianópolis (where the authors live). References to other Brazilian cities will be made when appropriate.

Brazil, a country of contrasts[2]

Brazil is located in the central-eastern part of South America, sharing boundaries with all the countries in the continent but Equador and Chile. Its population is 170 million (2000). It is a federate republic, comprising twenty-six states and a federal district where Brasília, its capital, is situated. The country occupies 8.5 million square kilometres of land and has an extensive Atlantic coastline. It is divided into five regions: Northern, Northeastern, Southeastern, Southern and Central-Western. These regions also form three large regional complexes known as Amazonia, Northeastern and Central-South.

Brazilian Amazonia – which consists of the entire Northern region and portions of the Northeastern and Central-Western regions – is dominated by an exuberant rainforest environment, inhabited from time immemorial by hundreds of Amerindian societies and gaining global significance from the 1970s. Its demographic density is low, having only two large cities: Manaus and Belém. The main economic activities in Amazonia are the exploitation of the Brazil nut, cultivation of rubber trees, the mining of manganese ore and casserite (tin ore), and agribusiness.

The Northeastern complex comprises the Northeastern region proper and the north of the Southeastern region. It is characterized by a semi-arid climate, severe droughts, extensive migration southwards and large rural properties, besides exhibiting strong indexes of poverty. It is the second regional complex in population density, hosting three large cities: Salvador, Recife and Fortaleza. It was the first region to be occupied by the Europeans.

The Central-South consists of the South and Southeastern regions and a large portion of the Central-Western one. It embodies the most important part of the Brazilian economy. More than 60 per cent of the country's population is concentrated in this region, which has five large cities: São Paulo, Rio de Janeiro, Belo Horizonte, Porto Alegre and Curitiba. Florianópolis is also located here.

During the colonial period (1500–1808), the foundation of Brazilian society was based on slavery, with the polygynous miscegenation of – that is, the plural intermarriage or interbreeding between – Portuguese men and enslaved African and indigenous women playing a crucial role in this formation. The abolition of African slavery in Brazil dates back to 1888, whereas indigenous Indian societies continued to be destroyed, enslaved or pushed further into the hinterland even after this date. Its economy was of an 'agroexporter' type, producing a large variety of crops and breeding many types of livestock, based on *latifundium*. *Latifundium* was a feudal system specific to South America that granted the *hacienda* (that is, estate) owner – the *latifundista* – an authority overriding written laws and placed every individual, even civil servants, in bondage to the landed estate of such feudal lords.

Industrialization took place in the Southeastern region and it attracted a relatively large numbers of Italian, German, Japanese, Polish and Lebanese immigrants to São Paulo and the Southeastern states. The Brazilian economy, totalling about US$900 billion (1996), exhibits one of the worst models of income distribution in the world. The Central-Southern region, the focus of this research, accounts for more than 50 per cent of the GNP, with only 36 per cent belonging to São Paulo.

Brazil is seventy-fourth in line, according to the UN indicators ranking human development (IDH) which measure a country's development on the basis of life expectancy, educational levels and per capita income. According to these rankings, Canada has first place and the USA is third. Brazil has a murder rate of 19.4 per 100,000 inhabitants, the third highest in Latin America. It is exceeded only by Colombia (78.6) and Bolivia (23.3). For comparison, the rate in Canada is 2.0 and in the USA it is 9.0 per 100,000.

A newspaper reading session

Reading the newspapers on a particular day can exemplify the diversity and intensity of urban violence in Brazil that is closely related with economic class, space and ethnicity, as will be explained later. We will take as an example the 'Daily Life' section from the 18 November 2000 issue of *Folha de São Paulo*, the newspaper with the widest circulation in the country. Each day this section covers the so-called *faits divers*, also known as the 'News in Brief' column. On this particular day we find news on different types of recurrent violence in the country. The whole of the first page is devoted to a massacre in the state of São Paulo, under the heading 'Ten die in the year's largest massacre in São Paulo.' Page three gives violence in Rio de Janeiro, under the title 'Army grenade explodes in Leme, wounding six'; another heading, 'Entrepreneur's son is freed', refers to the release of a hostage in São Paulo. In addition to this, we find on the same page – though with less prominence – news of a death in a prison cell during a fire; one dead and seventeen wounded in an mass attempt to escape from a penal colony; a note on two policemen indicted by the court for torture; the arrest of a doctor who murdered his wife; and a cyclist who was killed by a steamroller.

Thus, in this particular issue of the newspaper, chosen at random, we find examples of the urban violence that occurs on a day-to-day basis in Brazil: the slaughter of São Paulo slum-dwellers (which was not committed by a group of police, as is usually the case,[3] but by a rival group of dealers), the turf war of drug dealers in the city of Rio, the kidnapping of a member of the elite, the death of criminals in prison, male domestic violence and violence related to motor transit. It is necessary to ponder and comment on these examples of violence in order to understand the psyche and the involvement of the Brazilians, from various ethnic backgrounds, in urban violence and their use of space in this respect. It is equally important to keep in mind the fact that ethnicity or a culturally determined range of 'racial' characteristics, from 'black' to 'mulatto' to 'white', runs parallel to the economic classification of, respectively, 'working class', 'middle class' and 'upper class' in Brazilian society, with a majority of the working class living in poverty-riddled *favelas*.[4]

We will examine here the first case, which took place in Jacareí, a city only 168 kilometres from São Paulo (quite close by Brazilian standards).

> Ten people between the ages of thirteen and twenty-six were shot to death while asleep at home. The woman who owned the house had gone out early to her job at a bakery, leaving her children and some friends asleep. The house was broken into and all who were there were killed, victims of the more than one hundred shots that had been fired. According to the police, nine of the victims had already faced police charges of homicide, theft or drug trafficking. The police concluded that the incident was a settling of accounts among local gangs.

Massacres – mass murders – occur daily on the streets of many of the country's capital cities, most of them gaining little visibility since they take place in *favelas* late at night, far from the view of the middle class and the elite. Most of the time, they are orchestrated by the police or paramilitary groups that have been hired by local merchants anxious to rid the neighbourhood of thieves.

In truth, police impunity in Brazil is enormous. Cases in which police are indicted for torture (as covered in the newspaper article cited) are rare in a country in which torture is common practice in interrogating the poor. This has been amply documented in human rights reports. The torture of upper-class individuals occurred only during the military regime (1964–79); at the time, it was directed against left-wing activists.

In the case of the Jacareí slaughter, gang conflict is also involved – another form of crime that is commonplace in the country. Groups of drug dealers, fighting over their share in the drug-consumption market, engage in parallel wars that are followed by police from a distance.[5] This particular mass murder[6] took place in the outskirts of the city ('on a street with no asphalt', according to the newspaper) and involved white, mulatto and black youths. The number of shots fired – more than one hundred – illustrate a further point that is also evident in the piece entitled 'Army grenade explodes in Leme, wounding six': the ease with which bandits get access to arms and ammunition.

'Entrepreneur's son freed' shows a Brazilian invention, hostage-taking, which today has spread all over the globe. In fact, as the former *guerrilheiro* (guerrilla fighter) and current federal representative Fernando Gabeira (1980) tells us, this practice was invented with the kidnapping of the US ambassador, taken as a hostage to be used in exchange for political prisoners held by the military regime. Today, members of the poor and working classes commit the crime of taking hostages for ransom. Initially concentrating on millionaires (as is the situation in the case studied), the victim would be held hostage during a long negotiating period. Today, any person on the street may be taken hostage in kidnappings; the victim is often forced to take money out of an automatic teller machine in order to be released. These are the so-called 'lightning kidnappings' that occur by the hundreds in large Brazilian cities and are not reported in the newspapers. It is hard to get official statistics on kidnapping, since the victims, fearing for their lives and not believing in the police, frequently do not report to the police station. For cases like those reported in the paper cited above, involving the rich, the São Paulo and Rio de Janeiro police keep separate specialist anti-kidnapping squads.

The two following news items deal with prisons, the site of some of the greatest violence per square metre in the country. The subhuman living conditions in these locales has also been exhaustively denounced in reports to the UN Human Rights Commission. The most common prevailing accounts of their predicament are those of overcrowded cells in which prisoners have to sleep standing up; prisoners enslaved by their peers; prisoners forced to play a lottery to see who will be killed to leave more room for the others; and subjugation to police violence. The Carandirú massacre (São Paulo, 1996) is an extreme example of this type of violence. There the prisoners were barbarously slaughtered by police who invaded cells after getting a riot under control. As the Brazilian songwriter Caetano Veloso puts it in a song: 'They were 111 defenceless inmates, almost all black, or almost black. Think of Haiti: this is Haiti.' The police who were involved acted with impunity and remain in the police force today.

The last two pieces of news deal with domestic violence and traffic deaths. The visibility of these two types of violence occurs in inverse proportion to their statistical importance: more people die per year on the roads in Brazil than the number of North Americans who died in the Vietnam War.[7] However, it is important to note that the new and more severe national transit law passed several years ago has helped to reduce these numbers. With regard to domestic violence, present in all social classes (the murderer in the case described in this newspaper is a doctor), its targets are primarily women but also include children and the elderly.

Special police stations have been created to which battered women can report their complaints; their role has been emphasized in all studies on this type of violence. These studies have also shown women's preference for police action to stop men from using further violence against them, rather than the actual arrest of the batterer. Brazil has few shelters for battered women.

The home and the street

Today, primarily for the residents of large Brazilian cities, 'the home' and 'the street' are, respectively, paradigms of 'good' and 'evil'. 'The street' has become a peril, a place that has, paradoxically, become the residence of a significant portion of Brazil's low-income population, blacks and mulattos in particular.[8] 'The home', which once always kept its back door open for visitors, is barred off today, protected by alarms, anti-burglar systems, private security guards and video cameras. It can, in fact, be watched twenty-four hours a day through infra-red light surveillance devices that detect movement and send messages to the headquarters of private security firms. Devices attached to telephones identify the calling number and, as the most recent market invention, there are 'tele-voicers' that transmit a previously recorded message and indicate where the messages were set into action. Security equipment is, nonetheless, illustrative of a coming together of the 'old' and 'new' models of protection.[9] That is, we have on the one hand sophisticated state-of-the-art computerized equipment and on the other artifacts used with cruel intent, such as glass shards embedded on wall tops; iron fences topped with spearheads or spikes that are capable of terminally perforating or impaling a human body; plain barbed-wire fences; electric fences electrocuting the invader; or fences constructed from razor-sharp wire (sold in packages of twelve to sixty metres). The co-existence of these two models shows how – what Norbert Elias (1990) has called – the 'civilizing process' is produced, a movement which persists in contemporary societies.

The expansion of this paranoia can be measured by the growth in the country of an industry specializing in security which generated US$650 million in the year 2000, placing this sector on the list of industries with the highest growth in Brazil, at a rate of 25 to 30 per cent per year. Evidently, this is a market for the elite who are able to pay prices that are usually higher than the minimum monthly wage of a Brazilian worker. Entrepreneurs in the area estimate that the market for this kind of equipment includes three million residences, of which only 6 per cent are using their services today (*Folha de São Paulo* 16 October 2000). It should be kept in mind that, as a rule, apartment buildings, town houses and even residential streets in elite neighborhoods have security teams that guard the area.[10]

Residences are not the only target of this 'total defence' project. In fact, prior to the residential areas, workplaces, firms, banks and commerce in general had been covered by highly developed security systems. At present, automobiles have become a prime target, as another 'space' occupied by middle- and upper-class residents who never use the Brazilian public transportation system. Offers related to the protection of drivers range from bullet-proof windscreens to instruments with an almost invisible radio system from which, in the case of being held hostage, the victim can transmit discreet signals from the car's boot. One of the most successful devices offered during a recent automobile fair in São Paulo was the 'Mul-T-Lock' that, if installed in the car's central console, uses a steel bolt to lock the gearshift when the driver turns off the engine. Imported from Israel, the device costs US$150 in Brazil. Another piece of equipment manufactured by a Brazilian company is a car alarm using sensors that can block up to eight of a vehicle's vital functions. The number of armoured cars in the country doubled in the year 2000. The market for armoured private vehicles is also expanding.[11] The price of this equipment has come down but remains at around US$20,000 per vehicle, varying from the lighter armours (referred to as 'urban use') to those that can resist weapons of varying calibre and some explosives (referred to as 'anti-hijacking'). 'An armoured car is harder to drive, it's heavier', is how a sixty-year-old informant from the city of São Paulo describes the experience. Its excessive weight increases fuel consumption and can also lead to mechanical wear and tear in the car, since the vehicle was not originally made for this use. But this is apparently unimportant to the members of this 'culture of fear', who are even willing to drive army tanks. Even automobile companies, who have thrived on the obsession with protection that exists in the society, are offering for sale on the urban market vehicles intended for rural areas and unpaved roads. So the sales of so-called 'four-wheel drives', large Jeeps with traction on all four wheels, have sky-rocketed. These vehicles are expensive to buy and consume far more fuel than other automobiles, but they address both aesthetic and pragmatic notions of consumers.

There are reasons why cars are among the most protected goods in the country. Since they are in transit in public territories, they are perceived as vulnerable to attack. In fact, the hijacking of cars on city streets in broad daylight is not uncommon today. As in gangster films, one or two vehicles pull up alongside the targeted car, forcing it to stop even if this costs the life of the driver or the private security guards; the individual who will later have to pay a hefty ransom in order to be released is pulled out of the vehicle.

In addition to the physical protection of individuals and their patrimony, the protection of the automobile itself has become a grave concern. The number of car thefts in the country is quite high and vehicles are frequently taken out of the country, Paraguay being one of the prime destinations. It is also common for cars to be taken apart and sold as spare parts at black-market fairs that are clandestine but not secret – in fact they are well known and usually used by the very clients whose cars have been robbed! For this reason Brazilian car owners prefer and accept complete insurance for their car, which in other countries is thought of mainly as protection in case of an accident, as a necessary protection against the loss of the total capital invested in the car's purchase. In our interviews, theft was mentioned much more frequently than protection against traffic accidents.

The pager and mobile phones are other types of equipment whose use has increased greatly due to the culture of fear, serving as a shield for the individual in public spaces and particularly on the street.

> When my son goes out at night to a party or with friends, I get desperate. I call him all the time, to know where he is and if he's all right.

So said a fifty-year-old woman, one of the many mothers that uses a pager, when interviewed by a São Paulo newspaper reporter. The pager, through which the whereabouts of children of the upper class can be monitored, has of late been replaced by the mobile phone which has taken on the same function. Another informant, a forty-five-year-old woman and archaeologist who lives in Rio de Janeiro, explained:

> I call my friend the moment I get into my car and I talk to her all the way to my apartment, throughout the entire drive. When I lived in Santa Tereza[12] it was even worse. I had to move because I couldn't get home late, it was too risky.

The crisis of paranoia that has led the middle classes to protect their cars and their homes takes on a sinister light when it moves onto the streets. Visible and invisible barriers have been put up, not infrequently spilling over the borders of private space into public spaces.

The Truman Show

The most efficient formula for the protection of the street became popular as of the 1980s, with the success of the idea of the 'condominium' (an estate-like walled- or fenced-off town house complex). Formerly, there were the 'vertical condominiums' (apartment buildings); today there are also 'horizontal' ones (that is, the gated communities) that are private areas which can be quite extensive in size but function as an apartment building. In other words, they are organized according to internal rules decided on by the council of residents and, most of the time, limit access only to residents, their domestic workers and their guests.

Within these gated communities, the Brazilian ethnic cleavage is clearly demonstrated: on the one hand, white property owners participate in the meetings of the council of residents and, on the other, we find blacks and people of mixed race who provide services or are domestic workers. It is important to note that numerous individuals who would, in other countries, be placed into ethnically differentiated categories are all included into one category in Brazil. This has been the case for the descendants of Asians, such as the Japanese and the Koreans, or of Arabs and Jews, who live together harmoniously in these private spaces, all of whom identify totally with the hegemonic ethnic model in Brazil which considers them as whites.

In spite of the visible growth of a black middle class in Brazil, the latter rarely is able to reside in such condominiums, due to various mechanisms of social exclusion. 'A black's place is in the kitchen' is a well-known saying, which makes reference to the relegation of black people to the role of serving whites, as they continue to be associated in Brazilian culture with the enslavement of Africans during the Portuguese colonial period.

In almost all of the cities in Brazil possessing a middle- and upper-class population, these 'horizontal condominiums', alias gated communities, can be found, whether they are of the strictly residential type or the so-called 'resort' type that also reserves space for commerce and services. They create walled-in communities within cities, pulling down or pushing the actual walls or fences of their homes all the way to the perimeter walls of the area designated for the 'horizontal condominium' and implanting full-time surveillance, with watch posts, video cameras and armed security guards along these walls. All this is to ensure that there will be lesser surveillance and more freedom within these walls, so that the residents can live within a complex reminiscent of the set of *The Truman Show*.[13] This is an artificial reality based on the guarantee that poor and black people will and can only be admitted as employees. This social exclusion allows for older forms of sociability and residence to emerge within this ethnically and economically homogeneous space: homes without walls and fences, children riding bicycles down the street, women strolling along with their babies, couples out jogging. Sometimes there are cars patrolling the area, calming its residents with the constantly renewed sight of uniforms and arms.

The number of these gated communities has increased so much in large Brazilian cities that, for example, in São Paulo the local government recently decided to establish a special tax that their residents must pay, to be used in the Fund for Incentives to Public Security (*Folha de São Paulo* 27 December 2000). Thus every 'horizontal condominium' that has a specialized body of guards will have to collect a tax – meant to finance this new service – for the state.

Evidently, this new tax was not welcomed by the residents of these condominiums in São Paulo. One woman, who presides over the council of residents of such a condominium, protested:

> We're going to have to pay R$5,000 [US$2,500] to the state in order to have a team of guards when the state should provide us with this service.[14]

'We're going to sue them, not because of the amount, which is negligible [sic], but on principle', explained a lawyer, who believes that it is the responsibility of the state to defend the elite (*Folha de São Paulo* 29 December 2000).

It is interesting to think about the profound transformation of the experience of living in the city that has been brought about and consolidated by these gated communities: the experience of living in a 'bi-polar space', divided between the place of residence (the condominium) and places of leisure and shopping (clubs and shopping centres), each of them with restricted access and under full-time surveillance, and linked by the automobile which shuttles between the two. Much of what has been associated with life in the city – the *flâneur* of Baudelaire and Benjamin (1985) or, in Simmel's terms (1967 [1902]), the forced blasé indifference to so much stimulation – are hardly the probable attitudes for these residents. They go from one well-known space to the other inside a private vehicle meant to isolate its passengers as much as possible from what goes on in transit from one of these 'non-places' (Augé 1991) to the other, through barriers of varying types: sound (music), visual (smoked-glass windows), climatic (air conditioning) and physical (the car's armour).

The members of these new gated communities solve the problem of security through the purchase of 'inner avenues' within the complex, while purchasing their home and the impermeability they create on city streets. But what happens to the middle-class residents of the city? Those who are unable to move into these neighbourhoods that take after those seen in American movies and who still have to share their streets with the common people, continuing to experience the anonymity and the crowds that were so dear to those twentieth-century thinkers who reflected on urban environments.

The answer is, when the street is not private – as in gated communities – it must be privatized. This can be done in diverse ways that are visible within urban space and accepted by the authorities, although they are illegal in principle. The elite neighbourhoods of Rio de Janeiro are an example. A visitor passing through them, several years back, would have been shocked by the number of men in blue or black uniforms guarding entrances of the buildings in these neighbourhoods. They were the 'doormen', usually black or mixed-race emigres from the Brazilian northeast, who would only allow residents and authorized guests into the building. These 'doormen' continue to exist, but accompanying them are now armed guards contracted through private security firms, along with metal barricades that encroach on the sidewalks, making it hard for pedestrians to get around them. In some cases, such as the situation at the central plaza of Ipanema, one of Rio's elite neighbourhoods, a sidewalk running parallel to the regular city sidewalk was created when local merchants decided to unite their protective barriers. In other words, in order to get into the stores you have to be authorized by a guard who watches over the people entering into this elongated cage-like structure that fences off the entrance to the shops. On other streets, residents pooled resources in order to build bullet-proof watch posts that protect the guard who is on duty. The posts are similar to the guard booths in banks, though they differ inasmuch as they have been built on public ground that should be watched over by the state police.

Street security guards, uniformed 'doormen' and electronic devices similar to those that are used in the 'horizontal condominiums' are ways of maintaining social distance that keeps the white residents of 'vertical condominiums' away from their neighbours who live in the *favela*. In other major cities, barriers may be less subtle, such as the case of a four-kilometre wall built by the City Hall itself, in a neighbourhood of Salvador, with the intention of completely isolating the 10,000 residents of a *favela* (*Folha de São Paulo* 5 October 2000). In order to cross the street, the *favela*-dwellers must first walk for kilometres to get around the wall.

These people are the slum-dwellers that are seen as 'dangerous' by the state security apparatus but, paradoxically, they are the ones who are the most exposed to dangerous situations, as we saw in the newspaper items discussed at the beginning of this chapter. All of these strategies make up what has come to be called the 'culture of fear' that runs through different social classes. Fear of crime, albeit exaggerated, is not a fantasy but a real fact, since a large number of people have had to face some type of physical threat as victims. The fear felt by poor and working-class populations seems more visible and concrete, since this population is exposed not only to criminals (particularly drug dealers) but also to police harassment in the *favelas* when they come to search for criminals.

The unequal distribution of violence

Public and private security expenditure in Brazil exceeds what is spent on health. Yet this security expenditure is clearly concentrated in particular neighbourhoods of the cities and in particular regions of the country. With regard to regional distribution, it is well known that 81 per cent of all equipment is sold in the Central-Southern region (63 per cent in the Southeast and 18 per cent in the South) and 13 per cent in the Northeast. The Central-Western region and the North have the smallest shares of the market: 4 per cent and 2 per cent respectively (*Folha de São Paulo* 16 October 2000).

Although the security equipment and services market is concentrated in the richest part of the country and certainly in elite neighbourhoods, the same thing can not be said with regard to urban violence. Research has shown that urban violence is not uniformly distributed in all residential segments. Middle- and upper-class neighbourhoods have much lower levels of violence than the *favelas*. The city of São Paulo is a case in point: it is the richest major city and the one that has the highest rate of violence in the country – that is, an average of 66.89 homicides per 100,000 inhabitants (the average for the country is 26.8 homicides per 100,000 inhabitants).[15]

A map of homicide rates in São Paulo would show a clear concentration in the neighbourhoods located on the outskirts of the city – that is, where the number of *favelas* is greater. Although the areas of low homicide rates shrunk between 1996 and 1999, we can still find zones with a rate of less than 30 homicides per 100,000 inhabitants,[16] thus safer than many European cities. One such case is the neighbourhood of Jardim Paulista (a heavily forested neighbourhood close to the city centre and the residence of the two candidates in the recent mayoral race in the city). This contrasts with the slum areas on the outskirts of the city, where homicide rates are 90 per 100,000 inhabitants. It is also in these areas that the highest unemployment rates are found. While Jardim, one of the most centrally located neighbourhoods, has an unemployment rate of 8.3 per cent, the areas of 'the periphery' (that is, the outskirts) have rates as high as 21.2 per cent. This means that 71 per cent of the city's unemployed are concentrated in the outskirts. In some neighbourhoods, these numbers mean that one in every five economically active residents is unemployed, whereas in the centrally located neighbourhoods we find that on average there is one jobless person out of every 12.7 who are economically active. Almost half (48.25 per cent) of the unemployed are less than twenty-four years old, of whom 70.1 per cent live in the outlying zones.[17]

Similarly, in Rio de Janeiro there are close to 1.2 million people living in the *favelas* on the hillsides surrounding the city. This population has increased in recent years, growing from 15 per cent of the total number of inhabitants of the city in 1991 to 17 per cent in 1996, a growth rate along the lines of 7.93 per cent during a period in which the total population of the municipality grew by only 1.29 per cent.[18] The total number of *favelas* in Brazil is 3,905, having grown in the last ten years.[19] São Paulo is in first place, with 621 of them; Rio comes in second, with 513. Given their geographical proximity to the elite neighbourhoods, they have become a daily nightmare for the predominantly white middle- and upper-class population of Rio. The role of the police, as an ex-minister of security in the city stated, is to maintain a state of social apartheid 'without the need for the fences they use in South Africa, because they don't come down from the hills, they don't organize themselves.' The same ex-minister pointed to the low minimum wage (close to US$60.00 a month) as being what draws so many young people into drug dealing, an illegal activity that attracts young blacks and mulattos, and also drew attention to the police repression that leads to high rates of death.

> The guy sees his dad working his whole life for sixty dollars a month. He can make ten times as much dealing drugs. Of course he prefers drug dealing. It's a matter of logic. It's just not logical for someone who's never felt hunger.

Hot zones, cold zones

Today, some Brazilian cities – such as São Paulo – are monitoring urban violence by computer. Every complaint reported to the police is immediately put into an electronic system so that it is possible to find out at any police station the exact number of crimes reported in a particular city street or neighbourhood. This makes it possible to pinpoint dangerous or 'hot zones' on the urban map.

'Hot zones' include those that are considered potentially dangerous in many other cities of the world – poorly lit streets, dead ends, parks at night and so on – but there is also a long list of places that would seem inoffensive to an uninformed visitor. In most Brazilian cities, the areas around cathedrals, central plazas and bus stations are 'hot zones' that require precautions to be taken. Yet the most prominent 'hot zone' in Brazil today is to be found right on the street, in one particular place: at the traffic light. 'At night, the police themselves tell us not to stop at a red light. It's extremely dangerous', says a woman from Rio; this is just a piece of advice given to residents of São Paulo. Likewise, a magazine read by foreign residents in Brazil suggests to its readers:

> The risk of dying in a traffic light mugging is absurdly greater than in a hijacking. At night, calculate the time and speed you need in order to avoid getting caught at a red light. There have been no reported cases of assault to moving vehicles.
>
> (*Magazine of the American Chamber of Commerce*
> no. 367, December 2000/January 2001: 52)

Expressways, on the other hand, are considered 'cold zones' since the probability of mugging or hijacking is low, as the above-cited magazine suggests.

There is a word-of-mouth code that is known by residents of particular areas which guides people on what to do: don't stop at traffic lights; always travel with closed windows – even without air conditioning at a temperature of 40°C in Rio; avoid getting close to young street vendors (who could threaten the driver or passenger with razor blades); avoid wearing watches, brand-name clothes or training shoes, jewellery or other valuables in public places; and so on. These precepts are to be followed, even when doing so means breaking traffic laws or inconveniencing yourself or others. This code, known by big city residents, also obliges people to accept such unusual behaviour as always making sure there is cash in their wallet. It's commonplace in Rio to hear people to say, 'I only have money for the mugger' when they mean they don't have much cash. 'No one leaves home with an empty wallet', says the archaeologist cited earlier, since lack of money can irritate the mugger and put the victim's life at risk. Going to the beach, a habit held in high regard by Brazilians, also demands complying with precise rules. In Rio, for example, some luxury hotels keep security personnel on the beach, controlling the circulation of people there and thus preventing such common crimes as the theft of cameras. Newspaper and magazine reports also 'teach' Brazilian tourists from other parts of the country how to dress and what to take to the beach – in other words, what should make up the minimal kit of indispensable items.

'Hot zones' are not necessarily zones of complete anonymity, where bandits are unknown people. Several years ago we witnessed a mugging on a tram in Santa Tereza in Rio, in which a French woman tourist lost her camera to a mugger who calmly entered the tram carriage as it went slowly up the hill. The tourist resisted the attempted theft and, since the thief had tried to cut the camera strap with a penknife, he ended up cutting her. The mugger got off the tram as easily as he had got on, in plain sight of all the other passengers and passersby, who attempted neither to go after him nor to call the police. After helping the tourist, who was more panic-stricken than hurt, we spoke to the tram driver. He said he knew who the mugger was but would not report him to the police for fear of retaliation.

Thus, 'hot zones' frequently require a socially known code of conduct based on avoiding contact, due to the potential risk it implies. Among whites, contact with blacks and the poor is avoided; among blacks and mulattos, contact with police, drug dealers and rival gangs is avoided. When contact cannot be avoided, other socially known rules of contact are also put into effect in order to minimize the conflict between mugger and victim. One rule, to which the French tourist was apparently oblivious, is never to resist a mugging, even when it is going on in a public spot with witnesses present.

There are some Brazilian cities with low rates of violence, such as Florianópolis, 800 kilometres south of São Paulo. Nonetheless, here too we find all sorts of electronic security paraphernalia, as well as 'horizontal condominiums', security guards and so forth. This can be explained by the fact that many residents have moved to Florianópolis in an attempt to escape the violence of other large cities, so that they are already contaminated by the 'culture of fear'. Furthermore, these types of artifacts and places of residence are seen as markers of social distinction, symbols of modernity and wealth.

Gender, race and violence

This urban violence, which in the eyes of the majority of the residents of large Brazilian cities seems uncontrollable, is much more common, as we have seen, in poor neighbourhoods and slums inhabited primarily by blacks and people of mixed race than in the middle- and upper-class residential zones. Furthermore, there are particular groups of people who are considered more fragile than others and have become particular targets for violence: women,[20] the elderly, the physically handicapped, children and sex workers such as prostitutes and transvestites. These are groups who are the target of violence both in public spaces and within the private space of the family. Statistical data on victimization from the government's *Instituto Brasileiro de Geografia e Estatística* (IBGE) show that men die more frequently as a result of homicides in public spaces. These men are generally young, black or mulatto, slum residents and between the ages of thirteen and twenty. Thus women and men undergo different types of violence: women are beaten within the home and risk suffering rape on the streets. Young men, blacks and the poor are at greater risk of death on the street or before the age of thirty.

The global arms market and local consequences of violence

Statistics on violence in Brazil (and probably those for all of Latin America and eastern Europe as well) surprise Europeans. The high distributions tend to be seen as having been caused either by deep economic and social inequalities (in those analyses that emphasize the need for structural change) or as aberrations of a moral nature (thus emphasizing a need for more education). What rarely appears in these analyses is the role of the First World in maintaining such a state, whether through the exorbitant cost of financing foreign debt or, more directly, through the sale of arms to drug dealers. We will leave the first matter to the economists; with regard to the second, we see that Rio de Janeiro possesses and uses arms in ways seldom seen in other cities on this planet that are not engaged in war.

A large portion of these arms are in the hands of civilians, private guards and criminals. Evidence of this are the 251 grenades confiscated from drug dealers by the Rio de Janeiro police during the first eleven months of the year 2000 alone (*Folha de São Paulo* 10 December 2000). These grenades, according to the newspapers, are used against police stations and vehicles or in assaults; some are domestically produced, but a large number come from abroad. A documentary video by João Salles illustrates the global distribution of these armaments.[21] We see scenes of a huge room where millions of arms that police have taken from drug dealers are kept, while a police officer runs a commentary on the origin of the arms: they come from the USA, Russia, Switzerland, Israel, France, Germany and so on.[22] Just as in the case of the violence-augmenting arms, the security market too is completely dominated by foreign firms: 75 per cent of the security equipment sold in Brazil is imported. Of that share, half of it comes from the USA, 20 per cent from Israel, 20 per cent from Japan, 5 per cent from Canada and 5 per cent from other countries. In other words, both violence and the war against it produce profits in other countries.

Is the fear of the rich the same as the fear of the poor?

The street is considered deadly. That is why it is avoided by the predominantly white Brazilian middle and upper classes. They live within condominiums that are enclosed by walls and watched over, like medieval walled cities. From there they take their automobiles to other protected urban spaces, such as commercial centres or clubs. They are afraid to let their children out alone on the street, where their imported tennis shoes or bicycles can be stolen; they are afraid of being raped or kidnapped. Often the criminals are young thieves, hardly older than their victims.

Nonetheless, the street is even more of a deadly danger for the oppressed classes, made up of blacks and mulattos who live in slums where even the police will not go, where the law is dictated by gangs of criminals and drug dealers. People are perplexed when they learn that 'home, sweet home' is the site of domestic violence: violence against women, children and the elderly, as can be seen from the numerous complaints registered in specialist police stations.

To think of urban violence in Brazil as a uniform phenomenon would therefore be a mistake. Brazil is an extremely segmented society. The rich and the poor – that is, whites and blacks – live in two different but permanently articulated worlds, reflected in the urban space they occupy. The domestic workers (employed by most elite and middle-class families) become the connection or go-betweens for these two worlds (Velho 2001).

If we think of violence from the gender perspective, we see that the male population is more subject to violence on the street, while women are more subject to violence at home; men are more likely to die a violent death than women. If we think in terms of race, class and age, we have rich, white adult men at one end of the spectrum and poor, young, black men at the other. Life expectancy for these two groups is diametrically opposed. While the average life expectancy for the country is visibly on the rise, having increased by ten years in a very short span of time, the chances of survival for young, black males who live in slum areas are getting smaller and smaller.

Thus we see violence not as a fact unto itself, but as a profound indicator of the inequalities of Brazil and of the world. The struggle against police impunity seems, to us, to be one of the most important and difficult conflicts with the state in Brazil today, since it is within the state that violence is produced. The intermediaries are precisely those who are supposed to be involved in combating it. But the fear of violence of the civilian population is not new. The Brazilian composer Chico Buarque de Hollanda expressed this fear well in a song called 'Cry Burglar', set during the military dictatorship in the 1970s. 'Cry Burglar' describes a couple woken at night by alarming noises who fear that it is the police – a peril that is considered, by this song (and by slum dwellers), greater than that of a burglary.

Notes

1 These data were collected daily during two distinct periods: the first academic semester of 1995 and the second academic semester of 2000. The data collected in 1995 were part of a research project ('Project on violence against women and other minority groups') financed by the national research council CNPq with research teams coming from the NIGS/UFSC and the NEPEM/UnB.
2 The introductory information on Brazil has been taken liberally from Bastos (2000), supplemented with explanations of local terms for international readers.
3 The best-known was the murder of the children of Candelária Street, which led to the death of ten children and went down in history for having taken place in downtown Rio de Janeiro. It was perpetrated by police officers who were discontent at not having received the payment (a sort of protection money) that they were expecting from a group of street children who frequented the area. One of the survivors of this massacre recently furnished another spectacle of violence with inter-

national repercussions, in fact aired by CNN, when he hijacked a city bus in Rio and held several women at gunpoint for several hours. The police who got him under control later murdered him in their patrol car. They have not, as yet, been tried for the crime.

4 *Favela* is the term for the agglomeration of poor dwellings, illegally and irregularly built, which are usually set into the hillsides surrounding the city. The name came to be used around mid-century and comes from the name of the first of its kind, which emerged in 1906 on the Favela hill in Rio de Janeiro (Valladares 2000). The Portuguese term *periferia* (periphery) is commonly used for the same type of slum when not located on the hillsides.

5 In a documentary film, Hélio Luz (the ex-minister of security in Rio de Janeiro and, at present, a federal representative) recalls:

> One night, I saw the sky lit up with green lights back and forth between the hills [*favelas*] of Alemão and Santa Marta. It was a gang fight. I thought: 'They are using a weapon that was used during the Gulf War.' Now, what other country on this earth has a city in which gang fights involve that kind of weapon? Only in Rio.
> (João Moreira Salles, *História de uma guerra particular*
> (Story of a Private War), 1999)

6 In 1999 Jacareí, a city with about 170,000 inhabitants, was ranked twelfth for violent deaths among the cities of the state of São Paulo that had a population of between 100,000 and 199,000 (the murder rate was 42.9 murders per 100,000 inhabitants). An industrial city, it houses beverage makers (*Brahma, Kaiser* and *Pepsi*) and other manufacturers (*Parker, Votorantium Cellulose and Paper*). Data from the Fundação Estatal de Análise de Dados, cited in *Folha de São Paulo* 18 November 2000.

7 By October 2000, 150,000 people had met their death in traffic accidents on federal highways alone since the beginning of the year (*Globo News* 30 December 2000).

8 A recent incident in São Paulo illustrates the critical situation of homelessness in Brazil today. In a symbolic act organized to inaugurate the new city mayor, Marta Suplicy, a dinner party, not publicly announced, was held for leaders of the homeless peoples' movement. About 1,000 famished people showed up (*Folha de São Paulo* 3 January 2001).

9 'Security' is an indigenous category used to refer to private police. 'Security systems' refers to alarm systems and other electronic devices.

10 Given the low salaries that Brazilian police officers receive, a large number of private security guards are in fact police who hold two jobs: one as a state police officer and the other in private service as a security guard.

11 The engineers O'Gara, Hess and Eisenhardt created the first armoured car in 1942, during World War II, for use by the US president Franklin Roosevelt.

12 A middle-class neighbourhood in Rio that is considered dangerous because of its proximity to the Rocinha *favela*, the largest in Latin America. Rocinha has between 150,000 and 400,000 inhabitants and is one of the most developed of all such neighbourhoods: it has an Internet website that appears in four languages, a *McDonald's*, satellite antennas and so on.

13 A film about a television show that continues round the clock and in which Truman, the protagonist, lives on an island with hired extras – unbeknownst to him- that perform as his family members or friends, under total surveillance by TV cameras.

14 As reported on the *Globo News* programme, 28 December 2000.

15 Violence in São Paulo has been on the rise: in 1996 there was an average of 55.56 homicides per 100,000 inhabitants, increasing to 66.8 in 1999. This average increased in seventy-nine of the ninety-six districts of São Paulo between 1996 and 1999.

16 Data from the Instituto Nacional de Pesquisas Espaciais, cited in *Folha de São Paulo* 25 September 2000.
17 Data from the study by Barbosa, Alexandre *et al.*, *Mapa do emprego e do desemprego do município de São Paulo*, cited in *Folha de São Paulo* 4 November 2000.
18 Data from the city's annual statistics provided by the Rio de Janeiro City Hall and cited in *Folha de São Paulo* 21 November 2000. This document also shows that Rio has the second lowest annual population growth rate of all Brazilian capital cities, losing only to Belém (Northern region) which has a negative growth rate of 1.67 per cent. Rio's estimated population for this year is 5.6 million.
19 Data from the Instituto Brasileiro de Geografia e Estatística, *Folha de São Paulo* 7 January 2001.
20 The violence that women face remains hidden within the private and invisible space of the home. Ninety per cent of all reports of violence that are made by women in all of the states of the country – according to police reports – refer to spouse battering or a father's abuse (usually sexual) of a daughter. Silent violence – that is, domestic violence against women and children – takes place in more than 10 per cent of Brazilian homes, respecting neither social class, age nor ethnicity in its choice of victims.
21 João Salles, *História de uma guerra particular* (1999).
22 The same ex-minister of security to whom we referred earlier, Hélio Luz, proposes in his testimony that Latin American countries come together to control the arms factories in the countries from which the arms come:

> Just as they control coca plantations in Latin America, we should control and close down the factories in their countries that sell arms to the drug dealer. They won't sell arms to the IRA, to Saddam Hussein or Kadafi; we don't find those arms there, but we find them here!

References

Augé, Marc (1992) *Non-lieux: Introduction à une anthropologie de la surmodernité*, Paris: Seuil.
Bastos, Rafael (2000) 'Brazilian popular music: an anthropological introduction', *Primeira Mão* 40: 1–41.
Benjamin, Walter (1985) *Obras Escolhidas* (Scholarly Works), São Paulo: Brasiliense.
Elias, Norbert (1990) *O processo civilizador* (The Process of Civilization), Rio de Janeiro: Zahar.
Gabeira, Fernando (1980) *O que é isso: Companheiro*, Rio de Janeiro: Codecri.
Simmel, Georg (1967, orig. 1902) 'A metrópole e a vida mental' (On the City and Mental Life), trans. Otávio Velho, *O fenômeno urbano* (The Urban Phenomenon), Rio de Janeiro: Zahar.
Valladares, Lícia (2000) 'A gênese da favela carioca', *Revista Brasileira de Ciências Sociais* 15(44), October: 39–55.
Velho, Gilberto (2001) 'Biografia, trajetória e mediação', in Gilberto Velho and Karina Kuschinir, *Mediação, Cultura e Política*, Rio de Janeiro: Aeroplano, pp.3–12.

8 Ethnic consciousness arises on facing spatial threats to Philadelphia Chinatown

Jian Guan

Introduction

The Chinatown in Philadelphia, as other Chinatowns throughout the United States, is an area populated with mostly ethnic Chinese, marked with definite boundaries and located near central business districts in a metropolitan city. It was formed during the later nineteenth century and has struggled through the expansion of downtown Philadelphia, under the city's urban renewal and development projects, since the 1960s. Over the years, Philadelphia Chinatown has been pressurized and threatened; however, it has fought and survived. In the process, it has grown into a cohesive ethnic enclave, gained political and ethnic consciousness, and revitalized its distinctive culture.

Philadelphia Chinatown is located in the City Center, near the Delaware River and the Benjamin Franklin Bridge. It is adjacent to the Central Business District of the city and right beside The Gallery (the downtown mega shopping centre), the Pennsylvania Convention Center and Independent Mall; all of which took over land that for years previously had been inhabited in part by the residents of Chinatown. The Chinatown neighbourhood encompasses about twenty-seven city blocks bordered by Callowhill Street in the north, Arch Street in the south, 8th Street in the east and 13th Street in the west (Map 8.1). The heart of Chinatown is the major business area of 10th, Arch and Race Streets. The portal of Chinatown is the 'Friendship Gate' located at 10th and Arch Streets, built by Chinese architects from mainland China in 1983. Its authentic Chinese architecture represents Chinese cultural tradition and offers a symbolic welcome to those entering the territory of Chinatown. Before the Gate was built, the landmark of Chinatown was the tall building constructed in Chinese style on 10th Street. This building is now used by the Philadelphia Chinatown Chinese Cultural Center (a non-profit, charitable institution) and its parent organization Chinatown YMCA.

As an ethnic community, Philadelphia Chinatown represents both residential and commercial concentrations, serving the needs of ethnic Chinese and providing a major source of employment for its residents. It provides its residents and other Chinese visitors with a 'comfort zone', offering a sense of security and support in cultural, linguistic and spiritual aspects through social

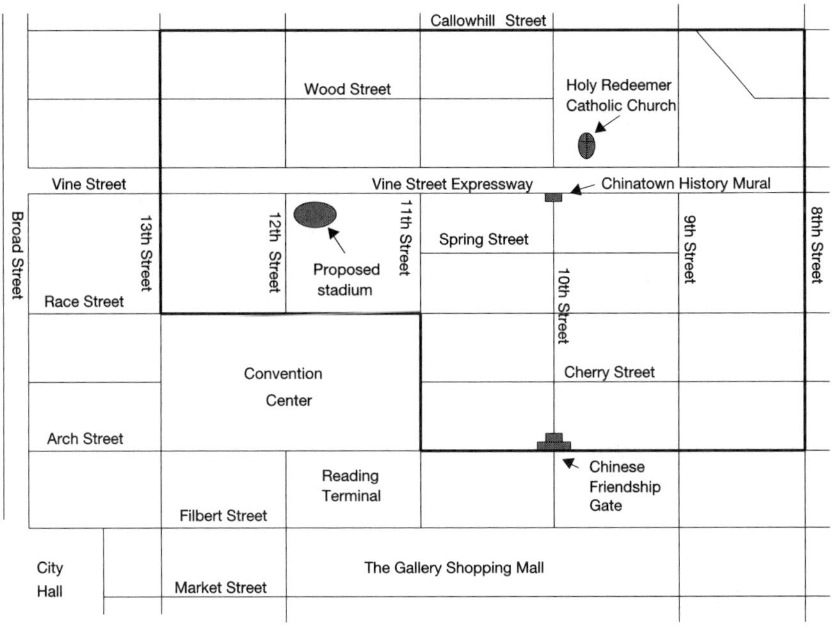

Map 8.1 Philadelphia Chinatown

organizations, churches and kinship associations. Today, Philadelphia Chinatown has been transformed from an ethnic neighbourhood to an economic, social, cultural and tourist centre that attracts both ethnic Chinese and American tourists. However, despite Chinatown's commercial prosperity, the people who live there are still ranked at the bottom of society in terms of socioeconomic status. As such, the land of Chinatown remains a target of speculation for city planners and urban developers.

The social status of Chinese people and the political position of Chinatowns in the United States have been shaped through a harsh historical process. Previous literature has contributed a great deal of understanding of the lives of Chinese immigrants (Chen 1980; Hsu 1971; Lee 1942, 1960; Miller 1969; Tsai 1986) and the history of Chinatowns (Gyory 1998; Ling 1998; Wong and Chan 1998). Other research has focused on the multiple functions of Chinese communities, along with the economic and cultural aspects of Chinatowns (Chow 1999; Light 1984; Lyman 1974; Wong 1982; Zhou 1992). Studies on ethnic identities have thrown considerable light on the Chinatown phenomenon in the United States (Kinkead 1992; Lyman 1986; Novak 1972; Vecoli 1995; Wong 1991; Wong and Hirschman 1983). Analysis of racial segregation and limited residential mobility explains the concentration of racial discrimination and poverty in Chinatowns (Zhou 1991). Chinatowns, as ethnic enclaves, are also experiencing change in the post-industrial era of globalization (Kwong 1996; Lin 1998). The Chinese in Philadelphia shared the lower socioeconomic status of their counterparts in the Chinatowns across the United States as the result of a history of institutional discrimination. Philadelphia Chinatown was formed in a pattern similar to the others, starting in a less than desirable location that was subsequently targeted for urban renewal and development projects (Aubitz 1988; Bock 1976; Cheng 1946; Jackson 1972; Jin 1997; Jung 1974; Lee 1976; Loh 1944).

However, while most research focused on historical discrimination in society and the voluntary segregation of Chinatowns, very few studies analysed the current phenomena of institutional discrimination and the vulnerable geographic position of Chinatowns in the face of urban gentrification in the United States. The current study utilizes the most recent information and uses the case of Philadelphia Chinatown to demonstrate how prevailing institutional discrimination sacrifices Chinatown to expansion and development. This research also explains, in relation to external pressure from mainstream society, the internal drive underpinning the socioeconomic development of Chinatown, through both the political consciousness created and Chinese cultural self-assertion.

Historical background

Philadelphia Chinatown was founded in the early nineteenth century. At the beginning of the century, the area was predominantly occupied by British home owners. By 1850, when the British moved westwards, Irish and German immigrants began to inhabit these streets. Commercial loft buildings and rows of houses replaced private housing. The area then started to deteriorate and, from about 1870 to 1890, it became known as the city's 'tenderloin' district, with burlesque theatres, hotels and rooming houses (*History of Chinatown* 1995).

Church records indicate that the Chinese population began living in this area as early as 1845. In 1870 there were approximately thirteen Chinese living in the area. Lee Fong opened the first laundry at 913 Race Street. A decade later, his cousin Mei Hsang Lou opened the first restaurant on the floor above this laundry. In response to the needs of the growing Chinese community, the restaurant gave people an opportunity to physically get together in one place. With the influx of more immigrants into this area, other laundries, restaurants and eventually grocery stores were opened by the Chinese. These Chinese people, predominantly men, often gathered at 9th and Race Streets, where they ate, played games and talked of their homeland and the New World. Since then, this area has become widely known as Philadelphia Chinatown (Cheng 1946; Culin 1891; Loh 1944). Based on US Census Office reports, by 1890 there were 738 Chinese in the city of Philadelphia. By 1900 there were 1,165 Chinese in the city, 276 of whom lived in the Chinatown area (Auyang 1978). The Great Depression, between the 1920s and early 1930s, caused a decline in the population of Philadelphia Chinatown, with many Chinese leaving for New York for greater job opportunities. World War II was the turning point for Chinese Americans who wished to integrate into American society. From

Philadelphia, 111 Chinese American men enlisted or were drafted into the US Armed Forces. As labour shortages cropped up in the war industry, more Chinese entered the American labour force. After World War II, especially after the 1965 Immigration Act, a new wave of immigrants moved into Chinatown. The arrival of Chinese women began to transform Chinatown from a 'bachelor society' into a family-oriented community (*History of Chinatown* 1995). Just as the neighbourhood started to expand – with more business and an influx of families in the mid-1960s – it encountered the expansion of downtown Philadelphia under city urban renewal and development projects. After these projects, the Chinese population in Chinatown dropped from 1,000 to 200 (Chen 1980). By 1980 there were 12,210 Chinese in the Philadelphia and Wilmington region (Tsai 1986), and about 600 residents and workers in Chinatown (Auyang 1978).

According to the US census of 1990 there were 1,198 Asians among the 2,485 people living in Philadelphia Chinatown, and in the 2000 census still no more than 1,229 out of its 2,422 residents were Asians (*Philadelphia Neighborhood Profiles* 2001). Informants from Chinatown admitted that the level of participation in the 2000 US census survey had not been satisfactory, even with encouragement from various community organizations and churches, since the number of ethnic Chinese living in Chinatown is usually accepted as being over 4,000. This is the number cited in publications from the Philadelphia Convention and Visitors Bureau (PCVB) and Asian Americans/Pacific Islanders in Philanthropy (AAPIP). Parallel to population growth nationally among Asians, the Chinese population has doubled during the last ten years. Refugees from Southeast Asia had begun, since the 1980s, to boost the population in Chinatown where the Asian population continues to grow. A tide of newcomers, mostly from China, occupy the available housing in the neighbourhood. The Chinatown community organizers speak of indicators pointing to an important concentration of ethnic Chinese in their community. For instance, one recent news report stated that at the Holy Redeemer School and Church, where almost half of the student body consists of newcomers from China, the enrolment had jumped from 150 to 246 over the last decade. The Chinese Christian Church at the heart of this neighbourhood had to add a Sunday service in Mandarin, China's official dialect, to accommodate over a hundred newcomers. In addition, more than 400 families are on a waiting list for subsidized housing (Lin 2000).

Ethnic enclave

An ethnic enclave is a residential cluster where members of an ethnic group live in the same place and share a common socioeconomic status and lifestyle. As an ethnic enclave, designated by the growing number of Chinese residents and Chinese businesses, Philadelphia Chinatown serves both its residents and its visitors. With ethnic Chinese making up over three-quarters of the residents of Chinatown and many Chinese living outside the complex, but working in and visiting the area, Chinatown makes a clear statement that it is a Chinese community.

As a visitor standing by the Friendship Gate on Arch Street, one finds a standard American business centre on the left and a remarkable Chinese shopping area on the right. Symbols of Chinese culture are visible in the architecture, as Chinese characters and decorations, in the open-stand grocery stores and hanging ducks at restaurant windows. On passing through the Gate, streets appear narrower, restaurants busier and grocery stores more crowded. Apart from those of the American visitors, white or black, the faces in Chinatown are overwhelmingly Chinese. During the weekend, there are more Americans and young Chinese, but on weekdays elderly Chinese are seen on the streets nodding to each other or chatting on the way to get their daily fresh food. Trucks for fish shops, grocery stores and vegetable stands manage to unload their products on the fully parked streets.

As a resident in Chinatown, an early immigrant or a newcomer, one can function quite well even without English as a communication tool. For instance, with only a knowledge of Chinese, people can find jobs in Chinese restaurants or grocery stores. They can find a place to live in the community by reading newspapers in Chinese or by word of mouth; they can take driving lessons and receive a licence and 'auto tag'; they can find a bank for their financial needs and get an accountant to fill in their annual tax return. All of these things can be done in Chinese. They can visit a doctor if they are sick; they can buy a phone card that provides services in Chinese if they want to talk to their Chinese friends or relatives in the home country, and can purchase aeroplane tickets if they need to fly home. They can get their hair cut, their clothes tailored and their bills paid. Again, these are all done in the Chinese language.

People who reside in Chinatown have a strong sense of belonging to the community. They clearly identify themselves as ethnic Chinese and with the territory of Chinatown. In their protests against the expansion of the highway project and that of the sports stadium, the people in Chinatown showed ethnic solidarity. For example, the gigantic mural 'History of Chinatown' by Arturo Ho, Giz, N. Phung and H. Tran (best viewed from where 10th Steet intersects with Vine Street) is the product of the 'Save Chinatown' movement. It expresses a sense of history and indicates the ethnic solidarity of the community there, as well as fighting back against the Vine Street Expressway running through Chinatown. Towards the bottom of the mural is an open book from which history 'pours out' along a right-hand 'page' that winds upwards, depicting scenes from the Chinese experience in the USA, before turning into a stream that runs towards the upper part of the mural. A giant laundryman – the founder of Chinatown – is wringing a wet cloth to produce the drops of water that create this stream. He has risen up from a mountainous horizon against the backdrop of the ocean and the sky. To the right of this unwinding 'page' is a stretch of railway track that ends in a tunnel at the craggy horizon, recalling the labour of earlier Chinese immigrants in the construction of the trans-American railway. Near the beginning of this 'page', in the foreground, is the palm of a big hand with the fingers stretched out as if to bring something to halt. This symbolizes the community's power to stop the expressway, as the the signs behind it – 'No Highway in Chinatown' and 'Save Chinatown' – make clear. Facing the Vine Street Expressway, the mural is the open declaration by a group of people who identify themselves with the territory of Chinatown. It is also a reponse to another mural in Philadelphia: the 'History of America', which depicts the 'discovery' of America and the arrival of the Pilgrims.

The major sources of employment in Chinatown have been in the service industry and manufacturing. The services provided in Chinatown can be categorised into four types. The first type basically serves the needs of the residents and helps the ethnic Chinese interact with mainstream society. These businesses include accounting, law firms, insurance, real estate, driving instruction, licence plates and travel agencies. The second type mainly provides services and goods for Chinatown residents and the Chinese population from surrounding areas, such as banks, doctors' offices, hardware stores, bookstores and video shops. The third type reflects Chinese traditions and can hardly be found outside Chinatown: Chinese herbal stores, acupuncture and martial arts houses can grouped in this category. The last type includes restaurants, bakeries, retail stores, beauty salons and special import stores. The latter two types of business serve a broad population, including Chinatown residents, the Chinese from suburban areas and tourists. Many of these businesses provide bilingual menus and staff for their customers.

Chinese organizations play an important role in preserving traditional culture, maintaining social networks and meeting the spiritual needs of the population. There are thirteen registered organizations. The Chinese Benevolent Association, On Leong Merchants Association, Hip Sing Association and other family-oriented organizations provide services for their members and new immigrants. Other organizations, such as Philadelphia Chinatown Development Corporation and Chinatown Parents Association, work with city officials to promote Chinatown as a viable residential and business community through zoning, housing and educational programmes.

The churches play a crucial role in education, in recreation and in providing connections with mainstream society, in addition to offering their spiritual services. There are three churches in Chinatown: the Chinese Christian Church and Center, the Holy Redeemer Chinese Catholic Church and School, and the Chinese Gospel Church. Because of the lack of public programmes in Chinatown, these churches have multiple purposes. They are dedicated to rendering services in the fields of religion, education, physical and moral development, social welfare and recreation. The Chinese Christian Church and Center focuses particularly on general community services. For example, it reaches out to new immigrants, helps with the paperwork for immigration purposes and provides professional services, such as health screening and accounting. The church is a major attraction for the young Chinese in Chinatown and for those from universities in the Philadelphia area. Members of various youth groups meet on different evenings and go to outings, picnics and annual retreats, as well as attending church services and Sunday School. The Holy Redeemer Chinese Catholic Church and School was established around 1940. Besides providing religious services, the Church operated as a school. In fact, it was the only school in Chinatown for many years. The school supplements its English courses with studies in Chinese history, language and literature. The church bus transports children between various sections of Chinatown and the school.

There is also a Buddhist temple in Chinatown. It has both elderly and young members. The temple provides regular chanting services for the safety and happiness of individuals and the community as a whole. It also organizes visits to surrounding temples and facilities, hosts national and international speakers, and distributes manuscripts and newsletters.

New immigrants from China (mostly from the province of Fujian) and Southeast Asia (mainly from Vietnam) continue to settle in Chinatown, which offers convenience and comfort to those with limited ability in the English language, poor knowledge of American society and an inability to interact effectively with mainstream society. As the interviews revealed, Chinatown remains a landing place for new immigrants and those who do not speak English. It is also a destination of choice for many Asian refugees, such as Vietnamese and Cambodians, after having been distributed from refugee camps. In order to stay in Chinatown, they are willing to pay the price outsiders will not pay. With over fifty restaurants, twenty manufacturing establishments and hundreds of businesses concentrated in a vanishing space, Chinatown is running out of room for commercial and residential development.

Struggle for living space

Even though the shopping strips and crowded residences in Chinatown are at the bottom of the social hierarchy, the real estate in Chinatown remains expensive and desirable because of the area's increasing population and the city's expansion into the area. However, this much-wanted area is shut off on all sides by impenetrable borders that prevent geographic expansion or outward development. Since 1964 Chinatown has been reduced to its current size by a series of city renewal and development projects. These projects include the Independence Mall Renewal Area Parking Garage in the east of 9th Street, a commuter rail tunnel and the Convention Center in the west, a downtown shopping mall in the south and Vine Street Expressway in the north (see Map 8.1). Chinatown has lost all its housing from 7th Street to 9th Street, as well as the lot on which the Convention Center was constructed. The Philadelphia Police Headquarters, Temple University's Institute of Feet and Ankles, Bell Atlantic Electronic Company, museums and the Tourist Center have replaced former Chinese residential neighbourhoods. It is fair to say that Chinatown has been deliberately targeted with regard to neither its right to self-determination nor the needs of its community. However, the people in Chinatown have never given up fighting for their living space.

The first organized protest occurred in 1966, when the project of converting Chinatown's major street – that is, Vine Street – into a six-lane expressway leading to the Benjamin Franklin Bridge was proposed. The residents of Chinatown protested against this project because of the anticipated noise from traffic and the threat to the existence of the Holy Redeemer Catholic Church and School – the only elementary school in Chinatown at the time. A 'Save Chinatown Movement' developed in the Chinese community. As the pamphlet *Philadelphia's Chinatown* (1998) put it: 'The once voiceless Chinese became intensely involved in the American political process.'

An influx of progressive, young Chinese-American professionals from suburban Philadelphia, along with the residents of Chinatown, became involved in the community's struggle for survival. The former have often been referred to as 'ABCs' (America-born Chinese): their parents possibly worked in Chinatown and they may have grown up in Chinatown themselves; they were regular visitors to Chinatown, possibly involved in Chinatown organizations or attending the church there. Now they utilized their knowledge of working with the media and their skills in public relations to help the community with which they were still identified. They were able to push Chinatown into using political force, including legal action, in its struggles for survival. For example, Cecilia Yep – a teacher, as well as being a young widow and mother of three – was the first woman to speak out in a public forum in Chinatown and to talk with the City Council immediately after the expressway was proposed. With George Moy and others, she organized the Committee for the Advancement and Preservation of the Chinese Community, incorporated as the Philadelphia Chinatown Development Corporation (PCDC) in 1969, which represented Chinatown in matters of urban renewal and community development. After twenty years of negotiations, the community of Chinatown and the city of Philadelphia agreed on a modified plan in 1986 that spared the church and the community. The Vine Street Expressway went underground and a noise-barrier was built on the south side of the two-lane ground freeway.

PCDC also focused its struggle on the planning of housing for the community and achieved another major victory: the first comprehensive plan both to preserve space and to build more residences in Chinatown was approved by the City Council in 1975. This was achieved through community pressure and the persistence of the PCDC, as well as through widespread and organized support from other groups around the city. Groups testifying on the housing issue at City Council meetings included the Asian-American Council of Greater Philadelphia, the PCDC, the Holy Redeemer Chinese Christian Center, the Chinese Student and Alumni Association, Yellow Seeds, the Council of City-wide Community Organizations and the Housing Association of Delaware Valley. As one member of the PCDC testified in 1974 at a Community Development Block Grant hearing,

> Our people had to settle in an area no one else wanted, the tenderloin section of the city. This was due to discrimination. By our industry, we have

made it what it is today – an asset to Philadelphia. However, times have changed, and our area is no longer undesirable. Large portions have already been taken for urban renewal projects ... . The center city plans east of Broad Street would have meant Chinatown's removal.

At the end of 1975 Chinatown won a major victory and the PCDC's housing plan was approved by the Philadelphia City Council. Some $2.4 million of city and state funds were committed to Chinatown for street improvements, acquisition and clearance of sites for housing (Auyang 1978).

Since then, the streets in Chinatown have been significantly improved and cultural elements were taken into consideration during that improvement. Bilingual street signs, Chinese symbols on the sidewalks and phone booths decorated with pagodas were among the elements preserved or introduced. However, the major achievement was the establishment of five major housing projects that included the construction of 230 residential units and 22 commercial units. Nevertheless, even with all this housing development, the overcrowded living conditions in Chinatown have not yet ameliorated. Slum housing in the back streets and 'shop-houses' (with shopfronts on the ground floor and families and employees living above) forced many Chinese to leave Chinatown for areas with more living space. Contributing to the process of suburbanization in America, an increasing number of Chinese have moved out of Philadelphia City to live in the immediate surroundings, while still working and doing business in Chinatown. For example, one of the informants, from New Jersey but a member of a Christian church in Chinatown, said:

> I drive to Chinatown only once a week, but my wife comes to Chinatown six days a week to teach at the kindergarten in the church.

His wife was a Harvard graduate with a Masters degree in education. Teaching at a public school close to her home would be more convenient and bring her more income, but she chose to drive to Chinatown for more than two hours a day because she saw the needs of the church and the children there.

Many of the Chinese have found employment in professions and industries in the suburbs where they live, but still drive back to Chinatown for weekends and holiday activities. One of these was another informant for this study, a physician who graduated from the University of Pennsylvania and started his business in a Philadelphia suburb twenty years ago. He drives three generations of his family to Philadelphia to attend church services. He, his wife and his four college-age children attend the service in English, while his mother-in-law goes to services in Cantonese. After attending church, his wife buys groceries and his children join their own groups. His daughter has joined the Church Chorus and one of his sons teaches at the Sunday School. The physician's children then have lunch with their respective friends in Chinese restaurants. The rest of the family stays on for the church meal, with important issues being dicussed at the lunch table. The grandmother socializes with an elderly group, while the physician prefers to participate, formally or informally, in decision-making on issues of importance to the church. As he put it,

> We had tried to attend nearby church for several years, but my children could not find their role models and we could not get involved. Now driving to Philly is a whole day event. It is time consuming, but the kids look forward to coming in every Sunday. It is rewarding.

Chinatown remains important to the Chinese and other Asian groups in Philadelphia. It is a cultural centre where traditional culture is preserved and ethnic identity perpetuated. It is a marketplace for Chinese food and other Asian products, as well as a meeting place for friends and relatives and a home for newly arrived immigrants.

The increasing need for community development forced Chinatown to look for new land. With limitations in every direction, as described earlier, Chinatown started its organized expansion by crossing over the Expressway to the north of Vine Street, reconstructing on the land covered with a ragtag group of warehouses and a neglected row of houses. In early 1998 a cluster of town-houses called *Hing Wah Yuen* (Prosperous Chinese Garden) was completed. It comprised fifty-one units of affordable housing for first-time homebuyers. These houses were fully occupied within a few months, with many people on the waiting list. Further projects for Chinese and other Asian business and residential settlements have been introduced, including the rezoning of approximately forty-four acres of currently under-utilized land which will further stimulate community reinvestment and allow a more rapid expansion of Chinatown to the north of Vine Street.

Political consciousness

Just as Philadelphia Chinatown began its development towards the north, the construction of a downtown baseball stadium at 12th and Vine Streets was proposed. The site was a mixture of an estimated 100 residential, industrial and small business properties. Chinatown leaders immediately criticized the idea and pointed out that the new stadium would cause more traffic congestion in an already over-crowded Chinatown and thwart plans for further development for the rapidly growing downtown Asian community. People marched in protest through downtown Philadelphia, following a business shutdown by Chinatown merchants (Lin-Liu 2000). People in Chinatown were also more experienced and more organized than before. There were four features that demonstrated their growing ethnic consciousness and improving political strategies compared to previous experiences.

First, the campaign was better organized. A day after the Mayor of Philadelphia announced the new stadium, leaders in Chinatown held a news conference at the Holy Redeemer Chinese Catholic Church and School – just two blocks from the proposed site of the stadium. They expressed concerns that the new stadium would further hem in a community already confined by city renewal and development projects. They stated that no one from Chinatown was on the stadium study committee, nor did the committee interview Chinatown leaders when it was considering several different sites for the stadium. They also pointed out that racial discrimination was at work with this plan. A petition opposing the Chinatown stadium was circulated and within a month the list had the names of 28,000 people.[1]

Second, the campaign was better planned. An umbrella group called 'Stadium Out of Chinatown', representing dozens of Chinatown associations, was quickly formed to organize the protests. It demonstrated Chinatown's anger and determination to oppose the plan. As Ping Leung Cheung, the co-chairman of the Chinese Benevolent Association, asserted: 'We've been kicked around for so long. This is the last straw.' Helen Gym, president of Asian Americans United, pledged that 'Chinatown is unified internally and externally in a way it has not been in a long time' (Chan 2000). The first protest was planned to coincide with a seminar in support of a downtown stadium, at which speakers likened Philadelphia to such cities as Cleveland, Denver and Baltimore, where new downtown stadiums had positive effects. To maximize the effects of the demonstrations, this protest took place at the same time as Chinatown merchants closed their shops. The people in Chinatown wanted their voice to be heard clearly this time.

Third, a broader alliance was built. During the preparation period, Chinatown merchants were approached, the suburban Chinese were informed by e-mail and friends were contacted. Thus, on the day of the protest, many of the protesters were neither Chinatown residents nor Chinese Americans. Several issues were used to build this alliance, including the higher taxes for schools built in Chinatown. The opposition from Chinatown started to gain sympathy among City Council members, who opted to lay the idea aside. The Chinese within and outside Chinatown had not simply waited for the final legislation to be introduced, they had created a network with the Chinese outside Chinatown who were better educated, wealthier and potentially more influential.

Finally, cultural resources were used to sharpen the issue's racial tinge. Protesters sat and knelt at the intersection of Vine and 12th Street, the site of the proposed stadium. The protesters marched through downtown beating drums and striking cymbals, while bilingual signs that read 'CLOSED to protest the Stadium' were hung in the windows of some of shops. Some held large metal sheets that they banged with their knees as the walked, making popping sounds similar to firecrackers (Lin-Liu 2000). They also passed by the Friendship Gate, an innovative mitigation measure by the City to preserve community cohesion. With this, they were not only trying to gain public attention, but also to clearly state that an ethnic minority group with a distinctive culture would be hurt by the stadium proposal.

In fact, what has happened in Philadelphia Chinatown is not an isolated incident. Chinatowns across the country are under attack, either from urban renewal – making the areas less affordable for the immigrant population – or from downtown development, such as new sports stadiums. For example, the Chinatown in Boston is at the mouth of an expressway tunnel. In Washington, D.C., the new MCI Center is swallowing up Chinatown. As Peter Kwong, a professor of Asian-American studies and the author of several books on Chinese immigrants, has said: 'Usually the least powerful political group is the Chinese. That leads to urban development projects dumped in Chinatown' (Brown 2000). In addition, Chinatowns usually cater for new immigrants, mostly unskilled workers and the immigrant elderly who do not speak English. When the new generation Chinese Americans acquire skills and better jobs, they move out of Chinatowns for a higher quality of life. This 'stepping stone' function makes Chinatowns necessary, but also vulnerable.

Cultural preservation

According to Vecoli (1995), group solidarity and ethnic identities are created, modified and reinvented in response to changing circumstances. So the greater the pressure from mainstream society, the greater the realization of cultural identity among the residents of Chinatown. This ethnic consciousness has been expressed in the process of rebuilding Chinatown, with many Chinese symbols being incorporated into its architecture. For example, the common character *shou*, symbolizing long life, can be found on the paving stones of the sidewalks, on the noise-barrier walls along the expressway and on buildings. Several murals in Chinatown catch the eyes of the visitor. As described earlier, the gigantic mural of 'History of Chinatown', facing the freeway to Benjamin Franklin Bridge, reminds residents and visitors of the history of Philadelphia Chinatown and the acts of resistance to the expressway. At the Gim San Plaza, another mural ('Vision of Paradise' by Chinese painter Lili Yeu) symbolizes the development and prosperity of Chinatown. In the playground of the new Chinese community *Shing Hua Yuan*, another product of the housing projects, there is a miniature railway track and train that informs the children of the Chinese history and contribution in the United States. There are also some clear outward manifestations of Chinese culture in Chinatown: for example, its major entrance, the Friendship Gate, with its authentic architecture; the bilingual (Chinese/English) street signs; and public telephone booths redesigned with pagodas on the roofs of the booths. Beautification is another theme in Chinatown. In an attempt to change the image of a ghetto-like Chinatown, many sites have been improved through planting trees, using decorated pots for decorative plants and creating gardens.

Chinatown people have also realized the importance of the media in cultural preservation and orienting the residents to the interests of Chinatown. Currently, there are five newspapers and two radio stations; two television channels are under construction. Learning Chinese is crucial for the children in Chinatown. In Philadelphia, 70 per cent of the 4,000 residents of Chinatown do not speak English but Chinese. Many of them find American mainstream schooling difficult for their children. Most parents of the 'ABCs' (American-born Chinese) are also worried about their children forgetting their mother-tongue. In response to this need, several Chinese language programmes are being offered by two private Chinese schools and three churches in Chinatown. The Chinatown Parents Association that emerged out of the 1994 campaign for a school bus for the children of Chinatown is now petitioning for a bilingual public school in Chinatown.

Philadelphia Chinatown serves not only its 4,000 residents, but also an estimated 250,000 Chinese Americans in Pennsylvania, New Jersey and Delaware. Debbie Wei explains:

> Chinatown is a symbolic center not just for the people who live there, but for the Chinese American community as a whole. What happens to Chinatown becomes a symbol for what happens to us as a community.
>
> (Brown 2000)

Wei is a regular Chinatown churchgoer and weekend shopper. As many others, she fears the cultural costs of building a stadium in the Chinatown neighbourhood. Though Chinatown plays a very import role in Chinese cultural preservation, it does not yet have a public school, a public library, a recreation centre or a park. When there is an urgent need for these things, the envisaged sports stadium is not a priority for the Chinese community, who are fearful of the negative effects such a complex would have on their culture.

Discussion of some observations

During its existence, Philadelphia Chinatown has witnessed a continuous push for internal growth and frequent threats of external infringement. Its struggle for space may never end. The attraction of Chinatown lies in its desirable location and amenities. For the ethnic Chinese, both inside and outside the territory, it serves multiple functions that are important to their physical and spiritual survival. For the municipal authorities and mainstream commercial circles, it is a natural target for urban development and renewal. Like its counterparts throughout the United States, Philadelphia Chinatown occupies a low power position in the administrative decision-making process of the city because of the low socioeconomic status of its residents, a well as being vulnerable due to its geographic proximity to the central business district. Also, the small size of the population in Chinatown makes it easier to sacrifice it for large urban development plans. Nevertheless, the people in Chinatown have protected themselves quite successfully up to now, as demonstrated by several residential development projects that have solidified the boundaries of Chinatown.

Chinatown is a segregated section in the city of Philadelphia. However, its existence goes beyond its physical space. Ethnic Chinese from the whole of Delaware Valley have continuously been involved in its struggle for survival. Successful suburban Chinese professionals who have a cultural consciousness and legal knowledge have taken an especially active role in supporting Philadelphia Chinatown financially, economically and politically. It is also fair to say that Chinatown belongs not only to the ethnic Chinese, but also to many other minority groups. For instance, many Vietnamese Americans consider Philadelphia Chinatown to be their home. There are several Vietnamese restaurants and a Vietnamese newsagent in Chinatown, which are meeting places for the majority of Vietnamese people. They have fought against the stadium alongside other people in Chinatown. Additionally, Chinatown has allies in mainstream society. Its current battle against the construction of the sports stadium in Chinatown will test the strength of such alliances.

As active actors, the people in Chinatown have been partial winners, using several successful forms of resistance. Forcing the expressway underground was very expensive and altering the location of the sports stadium was a hard decision. However, had it not been for the small size of the population and the lower socioeconomic status of people in Chinatown, they would not have become a target in the first place. While acknowledging that pressure from a community can affect the decisions of the City Council, we need to realize that the final decisions are not always based on the perseverance and willingness of the Chinese people, but on other considerations. In the case of the expressway, rather than claiming that the municipal administration was impressed by Chinese resistance, it would be more accurate to say that the administration was afraid to oppose the church. In the second situation, the decision not to locate the sports stadium in Chinatown was based on cost and accessibility rather than the benefits to the Chinese community. Building the stadium on land in Chinatown would have cost twice as much as in the other proposed location. In addition, Chinatown is already over-crowded and congested, without the additional traffic that would rush in and out of a new stadium. The municipal administration just could not generate enough avenues and add the necessary facilities to persuade the City Council to build the stadium in Chinatown.

Was there a happy ending? Yes, Chinatown won the fight. The Philadelphia City Council rejected the proposal to build the stadium in Chinatown. However, the decision was not based on the opposition they met with from the people in Chinatown, but on financial concerns: The cost of upgrading the Veterans Stadium in South Philadelphia was $395 million, rather than the $685 million it would have cost to build a new stadium in Chinatown. The upgraded sports facility is scheduled to open in August 2003, so there will not be a stadium in Chinatown now. The Chinese community would certainly support development projects that met Chinatown's urgent needs: a public school, a public library or a recreation centre, for example. But who can be sure that the next development project won't be at the expense of people of Chinatown once more?

Note

1   ABC Action News WPVI-TV Philadelphia, 'Philadelphia's baseball stadium', ABC Action News Special Reports, 14 June 2000.

References

Aubitz, Shawn (1988) *Chinese Immigration to Philadelphia*, Philadelphia: National Archives, Philadelphia Branch.

Auyang, Grace S.H. Chao (1978) 'Structural and processual change in Philadelphia's Chinatown and among suburban Chinese', unpublished Ph.D. thesis, Temple University, Philadelphia.

Bock, Deborah Lyn (1976) 'The historical function of Chinatown and its application to Philadelphia', unpublished M.A. thesis, University of Pennsylvania, Philadelphia.

Brown, Jennifer (2000) 'Philadelphia Chinatown's 30 years of struggling history', http://www.AsianWeek.com/, 22–8 June.

Chan, Anita (2000) 'Philadelphia mayor postpones release of Chinatown stadium plans to early fall', *aOnlin*, 27 June.

Chen, Jack (1980) *The Chinese of America*, San Francisco: Harper & Row.

Cheng Te-Chao (1946) 'Acculturation of the Chinese in the United States: a Philadelphia study', unpublished Ph.D. thesis, University of Pennsylvania, Philadelphia.

Chow, Julian (1999) 'Multiservice centers in Chinese American immigrant communities: practice principles and challenges', *Social Work* 44: 70–81.

Culin, Stewart (1891) 'Social organizations of the Chinese in America', *American Anthropologist* 4 (old series): 347–52.

Gyory, Andrew (1998) *Closing the Gate: Race, politics, and the Chinese exclusion act*, Chapel Hill: University of North Carolina Press.

History of Chinatown (1995) *History of Chinatown: 125th anniversary celebration Philadelphia Chinatown*, Philadelphia: Philadelphia Chinatown Development Corporation.

Hsu, Francis L.K. (1971) *The Challenge of the American Dream: The Chinese in the United States*, Belmont: Wadsworth Publishing.

Jackson, Catherine (1972) 'A study of the nature of leadership in the Chinese community in Philadelphia', unpublished M.A. thesis, Temple University, Philadelphia.

Jin, Dongzheng (1997) 'The sojourner's story: Philadelphia's Chinese immigrants, 1900–1925', unpublished Ph.D. thesis, Temple University, Philadelphia.

Jung, Marshall (1974) 'Organizational performance in an American Chinatown: an analysis of the criteria utilized by organizations in Philadelphia's Chinatown to assess their performance', unpublished Ph.D. thesis, University of Pennsylvania, Philadelphia.

Kinkead, Gwen (1992) *Chinatown: A portrait of a closed society*, New York: Harper Collins.

Kwong, Peter (1996) *The New Chinatown*, New York: Hill & Wang.

Lee, Rose Hum (1942) 'Chinese in the United States today: the war has changed their lives', *Survey Graphic* 31 (October): 419, 444.

—— (1960) *The Chinese in the United States of America*, Hong Kong: Hong Kong University Press.

Lee, Sharon (1976) 'Urban renewal: a study of Philadelphia Chinatown', unpublished paper.

Light, Ivan H. (1984) 'Immigrant and ethnic enterprise in North America', *Ethnic and Racial Studies* 7: 195–216.

Lin, Jan (1998) *Reconstructing Chinatown: Ethnic enclave, global change, globalization and community*, vol. 2, Minneapolis: University of Minnesota Press.

Lin, Jennifer (2000) 'Chinatown fears it will lose its way', *The Philadelphia Inquirer* 14 June: A01.

Ling, Huping (1998) *Surviving on the Gold Mountain: A history of Chinese American women and their lives*, New York: State University of New York Press.
Lin-Liu, Jen (2000) 'Philadelphia Chinatown threatened by gentrification', *Asian Week*, National and World Section, 22–8 June.
Loh, Chi-Chen (1944) 'Americans of Chinese ancestry in Philadelphia', unpublished Ph.D. thesis, University of Pennsylvania, Philadelphia.
Lyman, Stanford M. (1974) *Chinese Americans*, New York: Random House.
—— (1986) *Chinatown and Little Tokyo: Power, conflict, and community among Chinese and Japanese immigrants in America*, New York: Associated Faculty Press.
Miller, Stuart Creighton (1969) *The Unwelcome Immigrant: The American image of the Chinese, 1785–1822*, Berkeley: University of California Press.
Novak, Michael (1972) *The Rise of the Unmeltable Ethnics: Politics and culture in the seventies*, New York: Macmillan.
Philadelphia's Chinatown (1998) *Philadelphia's Chinatown*, Philadelphia: Balch Institute of Ethnic Studies, Balch Exhibit.
Philadelphia Neighborhood Profiles (2001) *Philadelphia Neighborhood Profiles*, Philadelphia: Philadelphia City Planning Commission.
Tsai, Shih-Shan Henry (1986) *The Chinese Experience in America*, Bloomington: Indiana University Press.
Vecoli, Rudolph J. (1995) 'Comment', *Journal of American Ethnic History* 14: 76–81.
Wong, Bernard P. (1982) *Chinatown: Economic adaptation and ethnic identity of the Chinese*, New York: Holt, Rinehart & Winston.
Wong, L. Ling-Chi (1991) 'Roots and changing identity of the Chinese in the United States', *Daedalus* 120(2): 181–206.
Wong, Kevin Scott and Chan, Sucheng (eds) (1998) *Claiming America: Constructing Chinese American identities during the exclusion era*, Philadelphia: Temple University Press.
Wong, Morrison G. and Hirschman, Charles (1983) 'The new Asian immigrants', in William McCready (ed.), *Culture, Ethnicity and Identity: Current issues in research*, New York: Academic Press.
Zhou, Min (1991) 'In and out of Chinatown: residential mobility and segregation of New York City's Chinese', *Social Forces* 70: 387–407.
—— (1992) *Chinatown: The socioeconomic potential of an urban enclave*, Philadelphia: Temple University Press.

9 Transcultural home identity across the Pacific

A case study of high-tech Taiwanese transnational communities in Hsinchu, Taiwan, and Silicon Valley, USA

Shenglin Chang

A Chinese Dragon is dancing, a band is playing music and a predominantly Asian crowd is gathering in large numbers. The time is not the Chinese New Year, nor the place Chinatown. In April 1997 *Ranch 99* had a grand opening ceremony for its fourth chain market at Milpitas Square in Silicon Valley, California. It was estimated that over 95 per cent of the crowd walking into the store were Asian consumers. Listening to the intonation of the dialect they spoke, it became evident that Taiwanese immigrant families dominated the Asian component. This is no surprise. *Ranch 99* is the name of a market chain specializing in Chinese food and home supplies. It is often the conduit for dozens of other Chinese restaurants, bookstores and groceries. In the advertisement for its grand opening, *Ranch 99* claimed that its Milpitas Square branch covered the largest floor space of all the northern American Asian shopping centres. What surprised me was how the Taiwanese immigrant families of the high-tech sector experienced *Ranch 99* in Silicon Valley and the way transcultural home identities and lifestyles emerged at the *Ranch 99* markets.

In Taiwan, the former chief planner of 'Hsinchu Science-based Industrial Park' (referred to hereafter as 'Hsinchu Science Park' or 'HSIP') told me:

> My husband and I live in Taiwan, but my parents and daughters live at Fremont, a city in the so-called 'Silicon Valley'. I fly between our Taiwan home to our Fremont home five times a year. Anyhow I don't quite feel the distance between Taiwan and Silicon Valley. The reason is simple. I can enjoy my perfect Taiwanese life in Silicon Valley these days, driving my mom to visit our friends in her Fremont neighborhood, shopping at *Ranch 99*, trimming and styling my hair in the hair salon at the *Ranch 99* Mall, etc. You know – those things you usually can only do in Taiwan. I also can buy all the Taiwanese groceries, plus enjoy very good Taiwanese food at *Ranch 99*.

In the suburbs of San Jose, Silicon Valley, USA, a 'high-tech housewife' – the housewife of a high-tech employee – explained to me:

There isn't any difference between Hsinchu and here, though my husband and I just moved here two years ago. You know, I run into some of my friends here at *Ranch 99* or the Chinese restaurants and I meet them in the [Hsinchu Science] Park when I fly back to Hsinchu. Though I have lived in the Park for more than a decade, America is a new country to me. I get really confused sometimes. I feel I have never left Taiwan …

Pondering on the intertwined relationship between *Ranch 99* and the Hsinchu Science Park, I interviewed fifty-eight high-tech personnel and 'high-tech housewives' in Silicon Valley, USA, and the Hsinchu Science Park, Taiwan. While it might be too early to predict how these two worlds – Silicon Valley and the Hsinchu Science Park – will intertwine in the future, it is possible to explore how transcultural home identities have emerged in high-tech centres across the Pacific. A 'high-tech housewife' told me that she felt as if she was living in Silicon Valley when she was in her Hsinchu Science Park home (that is, at Bamboo Village):

I lived in the United States more than ten years, mostly in the San Jose area, … but my family finally decided to come back here [the Hsinchu Science Park]. Though I have lived here for seven years, I don't quite feel I'm in Taiwan. I seldom leave home and the interior decorations of my home are the same as those in my previous home in the States. Except the space itself is much more tiny. My father-in-law always told me, 'Though it is very chaotic outside, as long as I enter your home, I feel I am in America.' I seldom walked outside, before I moved into the [Hsinchu Science] Park. The public spaces in Taiwan are hopeless. They are just too chaotic. Garbage, illegal parking, plus food stands that are everywhere. It is a lot of fun to be there [in Silicon Valley]. Many nice friends are there. Some of them are here, but I visit others there. I visit there every year. It won't be so much fun, if they are not there …

Along with the flood of high-tech professional immigrants into Silicon Valley, the interesting and striking phenomenon that emerges is that lifestyles are readily transported and intermixed, rather than being assimilated with the local cultures across the Pacific. 'Home identity' refers to an identification with what is considered to be 'home' or the definition of a home environment. Rather than the physical form of a house, a home is the dwelling with its surroundings which provides residents with the sense of 'home' and belonging. For some people, a home can be a house with a garden. For others, its boundary may extend to a neighbourhood, a community, a district or even a region. Therefore, an individual's 'home identity' not only represents a compartment of one's daily life, but also reinforces the social norms and values on which one builds when creating a dream home.

Transcultural home identity indicates that transcultural lifestyles have been merging into the homes of the people who are exposed to them. Instead of using the term 'multicultural lifestyle', I prefer to describe the new residential culture as a 'transcultural lifestyle'. The term 'transcultural lifestyle' is derived from the concept of transculturalism. There has been a theoretical breakthrough, with the initial concept of multiculturalism giving way to the evolving concept of transculturalism. Traditional concepts of cultures – including the classic single culture, interculturalism and multiculturalism – assume that 'every culture can be distinguished and remain separated from other folks' cultures' (Welsch 1999: 195). However, transculturalism 'sketches a different picture of the relation between cultures, not one of isolation and conflict, but one of entanglement, intermixing, and commonness' (Welsch 1999: 205). Welsch argues that we already live in a transcultural context.

> Lifestyles no longer end at the borders of national cultures, but go beyond these ... . There is no longer anything absolutely foreign ... . Today, in a culture's internal relations – among its different ways of life – there exists as much foreignness as in its external relations with other cultures.
> (Welsch 1999: 197–8)

Indeed, the concept of transculturalism has the capacity to analyse the transnational migration of Taiwanese 'high-tech families'.

'Transcultural home identities' establish a sense of 'transnational community', with the latter detached from the surrounding local communities. The 'transnational community' phenomenon refers to the way people combine their daily experiences, socio-spatial networks and place attachment to a 'counter-community' across national boundaries. Their individual lives are divided between two communities in two countries, while their community lives are similar in both places. Based on my survey, Taiwanese 'high-tech families' have developed their 'transnational community' lives because they connect with both Silicon Valley and the Hsinchu Science Park. They enjoy Silicon Valley, because they feel as if they are in Taiwan. They enjoy the Hsinchu Science Park community, because it is similar to Silicon Valley. I also assert that the transnational community phenomenon represents a restructuring of the relationship between the local and the global (Soja 1996; Tuan 1996). According to my study, the phenomenon consists of American-standard suburbanization and transcultural lifestyles in high-tech centres. It also proposes a new concept in defining Chinese immigrant communities (Anderson 1987, 1991; Fong 1994; Horton 1995). The essential questions for this chapter are: How has the 'transnational community' created by Taiwanese high-tech personnel disassociated itself from and/or associated itself with the local societies in both Taiwan and California? Investigating the landscape aspects of the phenomenon, how can we forecast any community transformations influenced by their respective transcultural home identities and lifestyles?

Restructuring Silicon Valley in Taiwan

Similar to the transplantation of the corporate campus and the office park development from the New York–Boston area to Palo Alto in the USA (Mozingo 2002), the planning concept behind Hsinchu Science Park has been a transplantation of a modified Stanford Industrial Park.[1] The Hsinchu Science Park is a combination of small-town community and industrial park development. It contains an industrial area, a residential area, a recreation area and service or commercial areas. According to Hsinchu Science Park (2001), 90,000 people commute to work there, while 2,000 both live and work there.

The residential area is known as 'Bamboo Village' and includes 645 households housing 960 individuals. The Bamboo Village in HSIP is a more self-sustaining community than most typical American suburbs. All of its residents are high-tech personnel who work in the HSIP and their families. Surrounding the village is a small supermarket, a post office, a bank, a petrol station, a clinic, a garage, a bilingual school, parks and open spaces, as well as a recreational area combining a swimming pool and tennis courts. Everything is within comfortable walking distance, very unlike suburban developments in Silicon Valley. In addition, there is an exclusive bilingual experimental school: only children who reside in Bamboo Village may enrol.

In order to recruit outstanding high-tech engineers from Silicon Valley, creating an American-standard quality of life and home environment was an important strategic goal during the planning of the Hsinchu Science Park (see Figure 9.1). Today, the Hsinchu Science Park is a symbol of Taiwan flourishing and influencing the high-tech global landscape in the information age. The 605-hectare Hsinchu Science Park is surrounded by a wall, causing it to be recognized as a 'unified kingdom' by 'outsiders', especially the local residents of Hsinchu. The landscape of the Hsinchu Science Park is indeed different from any 'urbanscape' in Taiwan. The well-maintained green spaces have become an important aspect of the Hsinchu Science Park in people's minds. The Hsinchu Science Park also has wide tree-lined boulevards and buildings surrounded by parking spaces. It is essentially car-oriented within its own boundaries.

How is the Hsinchu Science Park related to local culture and communities? The Hsinchu Science Park is located in Chin-shan-mian, a traditional Chinese religious location that has been centred on the Chin-shan Temple since the latter days of the Qing Dynasty. Before the Hsinchu Science Park was established, the main population in Chin-shan-mian had been the Hakanese, a subcultural and subethnic group of the Taiwanese.[2] Since 1979, being driven by high-tech policy, the Hsinchu Science Park has gradually acquired 605 hectares of land, mainly from the Chin-shan-mian area, and established the first high-tech science park in Taiwan. The industrial transformation boosted the 'backward' Chin-shan-mian farm village into becoming a 'global golden child' (Lee *et al.* 1997). This dramatic transition has not only generated billions in revenue, but also provided approximately 90,000 high-tech jobs in 2000 (Hsinchu Science Park 2001).

Figure 9.1 An American home environment in Hsinchu Science Park

During the development of the Hsinchu Science Park, the Hakanese local population in Chin-shan-mian village were relocated outside the Hsinchu Science Park boundaries. The relocation not only segregated the 'high-tech families' from the local Hakanese, but also changed the 'insider/outsider' relationship of the Hsinchu Science Park to its surrounding Chin-shan-mian area. New 'insiders' (the 'high-tech families' attracted by the Hsinchu Science Park) would eventually prefer to make their homes in either Taipei or Silicon Valley. They live inside the Park without any connection to the existing Hsinchu society.

'Parkers' or 'high-techers' are nicknames used for 'high-tech families' living in the Hsinchu Science Park. Everybody emphasizes that living inside the Hsinchu Science Park makes them feel as if they were in the USA; in the same way, the Bamboo Village reminds them of their Silicon Valley homes. 'Hsinchu region is nothing like Silicon Valley at all. The whole city has been growing without any plan,' one Parker remarked. They contrast the Hsinchu Science Park with Chin-shan-mian village and the Hsinchu region:

> Every time I drive into our Park, I take a deep breath, then feel so relaxed. To me, our Park is more like a place in America.

All agree that living in the Bamboo Village is very similar to their American residential experiences, for both spatial and social reasons. The spatial reasons are:

1 low residential density,
2 a secure living environment,

3   luxurious green spaces and
4   high-quality education facilities.

The social reasons are:

1   the similarity of social class and
2   shared transnational residential experiences.

In contrast, after the relocation related to the Hsinchu Science Park, the indigenous population under the jurisdiction of Hsinchu City and that of Chin-sha-mian became the 'outsiders' of the new Hsinchu Science Park. The transplanted Hsinchu Science Park changed the surroundings of their communities dramatically. The traditional Hakanese villages were demolished. Historical temples were relocated, disturbing the religious network,[3] while land-use patterns were transformed. Rice fields and sugar farms were eradicated, along with the traditional irrigation ponds, which were all filled in.[4] The local pottery industry declined (Lee *et al.* 1997). Instead, American-standard road systems were laid out and cutting-edge high-tech companies moved in, while quasi-American suburban housing design was adopted. According to local documents and information gathered during my interviews, the majority of the local residents were most concerned about three particular types of environmental impact. First, the Hsinchu Science Park Administration Office manages badly the toxic discharge from high-tech water usage. The local residents have protested to the Administration Office many times since 1997, but the office

Figure 9.2 Traffic congestion in Hsinchu Science Park

has seldom taken their petitions seriously. Second, the heavy traffic congestion has become a nightmare for the residents living adjacent to the Park (see Figure 9.2). It continues to grow worse. Third, the cost of living has escalated rapidly.

Intertwining 'here' and 'there' homes

Attracted to the jobs created by the Hsinchu Science Park, the population in the neighbourhoods adjacent to the Hsinchu Science Park and Hsinchu City grew rapidly. In spite of local high-tech employees emigrating from the Hsinchu region, two waves of reverse brain-drain from Silicon Valley in the USA brought over Taiwanese who had decided to work in the Hsinchu Science Park. The first wave moved into Bamboo Village in the early 1980s. They helped establish the foundation for the Hsinchu Science Park technology and economic performance. When they first moved back to Taiwan, Bamboo Village was a deserted ghost town. Some of them made their homes in Taipei City. Others had their families stay in Silicon Valley in the USA. The former group became the so-called '3–5' group, meaning they shuttled between Hsinchu and Taipei every Wednesday (the third day of the working week) and Friday (the fifth day). The latter group, with their families in Silicon Valley, became known as the 'astronaut families', because they flew back and forth between Silicon Valley and the Hsinchu Science Park. The second wave of the reverse brain-drain created a collective movement back home in the early 1990s. These families either set up home in the Americanized and walled-in communities of Hsinchu, or joined the high-tech first wave 'astronaut families'. The 'high-tech husband' would work in the Hsinchu Science Park, while his wife and children stayed in Silicon Valley, because the couple preferred to have their children benefit from the high-quality education available in Silicon Valley. The families, especially the wives, would fly back and forth between the two settlements many times during a year.

We can gain an insight into this transnational community by investigating the 'Housewives Club' network active in both the Hsinchu Science Park and Silicon Valley. The club functions as a foundation that connects the 'high-tech wives'. Core members of the club comprise more than 200 housewives. The club started very much as a grassroots and face-to-face process, in the form of daily chats in a public foyer. These women wished to support each other in the education of their children, daily chores and family-based recreation. They exchange their American experiences and update information on the education of their children at formal and informal gatherings, either in the Hsinchu Science Park or in Silicon Valley (Chang 2000; Tseng 1995: 41–54). The transnational network of these housewives now extends even beyond the Hsinchu region and Silicon Valley, having become a global network. A founding member explained how she had been upset by the unavoidable and frequent separations from her good friends:

> Many of our members fly back and forth between the Park and Silicon Valley, but Silicon Valley is not the only place our members migrate to.

Some of them migrated to another country for a couple of years and then came back to our Park for business reasons. Twenty-three of my twenty-seven neighbours have emigrated to other countries. They mainly moved to the United States, Canada, Australia, New Zealand and mainland China. I suffered from this kind of separation before. I got used to it gradually, since the engineer families in the Park have moved in and out frequently. I also realized that members of our housewives club could hold parties in Silicon Valley, when I visited my friends there last summer.[5]

However, while their transnational community relationship links the restructured communities in these two places, it also segregates the members from local communities in both places. Few of them have social networks with local citizens.[6] Their intimate neighbourhood network does not extend to Chin-shan-mian or Hsinchu City, but connects to a global Hsinchu Science Park community. The impression that local residents have of them reflects, to some extent, the experience that the 'high-tech families' inside the Park have lived for years:

There are many 'green card moms'[7] inside the Park. They don't need to work, but just fly back and forth between Hsinchu and Silicon Valley for two reasons. Firstly, their husbands can support them. Secondly, their children are studying in America and they are currently applying for their 'PR' (permanent residential status of the USA). Since they regularly fly back and forth between countries, no company will hire them anyway. They have their housewives club. It's a very exclusive club. They have their upper-middle class activities – very Americanized.

Although criticized by local residents, the housewives club is the most important family-based social network inside the Hsinchu Science Park. Not all of the members are 'green card moms', but all are residents of the Hsinchu Science Park and have travelled extensively around the world. They have no intention of extending their network to the locals in Hsinchu City, the Chin-shan-mian community or their neighbours in Silicon Valley.

Actions for the transcultural life

There are four types of collective action taken by the 'high-tech astronaut families' living in Bamboo Village:

1   recycling and gardening to develop environmental awareness;
2   'sporting' as a form of recreation that builds a high-tech class identity;
3   shopping for contemporary fashion; and
4   'shuttling' for children's activities.

All of these actions are carried out on a global scale.

First, with regard to the recycling and gardening activities, my interviewee (another core member of the housewives club) declared that she learned these during her year in Palo Alto. She appreciated the information and activities provided by her Community Center. She thought it was important for their neighbourhood in Bamboo Village to have the same activities as those in Palo Alto. However, she complained that other local residents outside the Hsinchu Science Park never paid attention to the public landscape and recycling issues. Even worse, she declared that local neighbours brought their rubbish bags and dumped them in her rubbish bins. She recalled her Silicon Valley experience:

> For example, the Community Center programs were held on a weekly basis, info on how to recycle garbage, etc. I felt it was great to have a community like this. The lives of 'high-tech housewives' were very happy and joyful there [in Silicon Valley]. We hung out together and chatted, but did nothing. However, our husbands had to work very, very hard. Nevertheless, I still often ask my husband if we can move back to Silicon Valley, or if we can migrate there.

Second, 'sporting' (that is, taking part in sports), as a form of recreation to achieve and maintain a high-tech class identification, has become a family phenomenon. Top-level managers and CEOs – the 'husband' group – play golf both in the Hsinchu region and in Silicon Valley. Playing golf not only manifests class identity, but also embraces Taiwanese high-tech pride along with high-tech business success. One informant told me:

> When I play golf in Silicon Valley with those Americans, I like to speak Chinese. You know, they are the ones who want to do business with me. Why should I speak English?

They also bring their children to practise. A housewife told me:

> My son is only a teenager, but he enjoys playing golf with his father. They play golf together every two weeks. It only cost NT$6,000 to NT$8,000 each time.[8] It's a long-term investment for him, so I don't think it's expensive.

Third, travelling is sometimes combined with 'shuttling' for the education of the children. 'High-tech families' prefer taking vacations in foreign countries. Many of them have friends and relatives living in other countries. According to my interviewees, the typical sequence for foreign travel comprises two steps. First, 'high-tech mothers' bring their children to American summer camps. They may stay with their friends or relatives for some time during this period. Afterwards, 'high-tech fathers' take about two weeks' vacation to come and meet them. They then travel to other places together. One of my interviewees said:

> Every summer, my sister's home is like a hotel. There are always some teenagers staying with her. These teenagers are either her relatives, or are from her friends' families. They come here [to Silicon Valley] to attend summer camps. If I cannot accompany my kids next summer, I will send them to my sister, too.

Another interviewee asked if I could go on an Alaska cruise with her family after her kids' summer camp came to an end.[9]

Finally, shopping with an international taste is the most obvious activity of the 'high-tech wives'. My housewife interviewees told me that they seldom shopped in Hsinchu City. Their shopping habits are still very American. For daily groceries, they prefer going to the recently opened megastores located near motorway exits. However, despite the efforts of local developers to make the shopping centres look 'Californian' or similar to such centres other popular places in the world, the 'high-tech families' are not satisfied. For high-quality home decorations, they would rather go shopping in Silicon Valley or Taipei:

> There are not many good stores to get the stuff I really want, so I go shopping in the Bay Area every summer, when I bring my kids to attend summer camp there.

I also conducted a group interview with eight 'high-tech wives' from the Silicon Valley. They all belonged to the Hsinchu Science Park housewives club and were active in the 'shopping' phenomenon. Their conversations were dominated by such topics as how to get into a good school district in Cupertino;[10] where to buy cheap brand-name clothes; how to get a good weight-loss diet plan; and who were the best members with whom to play *Ma-chiang* (mahjong). *Ranch 99* acted as the place where they could gather for their daily shopping and social life. In Silicon Valley, they usually met at the home of one of the members and chatted for a while. Then they would go to a nearby *Ranch 99* to have lunch. After lunch, they might go back to that person's home and have tea. Or they might go to another friend's place to play mahjong together. During this period, some of them might come and go, picking up their children or doing other things. Then all of them would go back to their homes in the late afternoon to prepare dinner before their husbands returned.

The transnational *Ranch 99* syndrome

On reviewing the transnational network of these housewives, we are led to wonder at how the '*Ranch 99* syndrome' became emerged as a cultural symbol for the Taiwanese employed in the high-tech sector in Silicon Valley.[11] The 'high-tech astronaut families' pointed out that they do not feel they are in the USA while staying in Silicon Valley. Six separate *Ranch 99* markets have opened in Silicon Valley within the past six years. *Ranch 99* provides everything they are accustomed to having in Taiwan. Furthermore, the *Ranch 99* chain supplies most of the trendy seasonal desserts from Taiwan. They can also rent recent TV series and popular karaoke DVDs or videotapes. Most interviewees responded positively, saying that they liked to go to *Ranch 99* immensely because they felt comfortable and had 'a sense of home' when they shopped there. The comparison made by some engineers between their *Ranch 99* experience and a visit to San Francisco's Chinatown, as given in my field notes, was:

> *Ranch 99* represents a much stronger identity than Chinatown. Comparing the new *Ranch 99* with the old San Francisco Chinatown, Jason states: 'Well, eating and drinking are really important activities in Chinese culture. *Ranch 99* may represent some collective identity for us under certain circumstances.' However, 'Chinatown is way too crowded and chaotic. For the traffic reason, I seldom go there. I do not like it. It looks like an enclave. The people who live there can hardly become members of mainstream American society. However, I do agree that it has cultivated many excellent second-generation Chinese Americans. Well, the problem is that they do not want to stay in Chinatown, either. Chinatown, to me, is a mother who never gets the nurturing from her kids. She has drained out her blood for feeding her kids.'

In fact, the immigrants in high-tech jobs seldom visit the San Francisco Chinatown. Many of them define Chinatown as a crowded, backward and chaotic enclave for poor immigrants from China, whereas the *Ranch 99* chain markets, taking the form of modern American megastores, more accurately symbolize the collective identity of Taiwanese immigrants employed in the high-tech Silicon Valley. One engineer explained his *Ranch 99* experience:

> I feel I'm back home at *Ranch 99*, especially the Great Mall in Milpitas. I don't feel I am a minority at all. I can buy all the things I need there. Although JC Penney is built in a similar American style, everything they sell at *Ranch 99* or the Great Mall is so familiar to me and is related to Taiwan. I will say the *Ranch 99* can represent my culture.

The architectural form of the *Ranch 99* markets is typical of suburban malls. Most Americans do not distinguish the *Ranch 99*s from other American malls. However, most Taiwanese immigrants, both high-tech employees and other groups, feel that *Ranch 99* symbolizes their collective Taiwanese identity. What do they mean by feeling 'a sense of home' at *Ranch 99*? Why do they feel *Ranch 99* is very similar to Taiwan? To which 'Taiwan' are they referring?

Interestingly, I found that the 'Taiwan' they refer to is the 'reconstructed' Taiwan in the Hsinchu Science Park. More precisely, the 'Taiwan' they define differs radically from the stereotype of San Francisco Chinatown. In my survey, they used the following words to describe the structural forms of *Ranch 99*s: 'modern', 'clean', 'well managed', 'spatially luxurious'. Therefore, the physical forms of the *Ranch 99*s are not at all similar to traditional Taiwanese commercial streets, community markets or even supermarkets. In terms of the interior operation of the markets, they explain 'a sense of home' as being intertwined with a predominantly Chinese environment and familiar, daily supplies from Taiwan. More importantly, one can not only shop for groceries here, but also have the alternative choice of authentic Chinese food, as well as newspapers, videotapes, music and books, besides being able to converse with friends at the same time – and all in one place. In other words, *Ranch 99* has become a quasi-Taiwanese Community Center in Silicon Valley.

In fact, the shopping experience at *Ranch 99*s differs from the typical street shopping or night market experience in Taiwan. There are no food pedlars outside the *Ranch 99* malls. Everything usually sold by street vendors in Taiwan is, at *Ranch 99*, nicely wrapped and beautifully displayed on the racks in the mall. One of the most obvious comparisons between the two places is the way that the most popular Taiwanese beverage, '*Po-ba* milk tea', is sold. In Taiwan, it is mostly sold by street vendors and costs less than US$1.00. In contrast, different shops sell the same drink at *Ranch 99* branches for at least US$2.50. Some stores have 'buy one, get one free' promotions, while others have monthly coupons (for example, 'buy eleven cups and get the twelfth for free'). The methods used to promote this drink are very Americanized, with the place to drink it being in a very modern mall. When my interviewees had the drink at *Ranch 99*s, they felt 'we are in Taiwan!' and 'this is so similar to Taiwan'. However, most of the high-tech employees that were interviewed complained about the dirty streets in Taiwan. They declared, 'I never visit street vendors in Taiwan'. The question is: What is the 'Taiwan' they refer to when they have their '*Po-ba* milk tea' at *Ranch 99*? The 'Taiwan' they refer to is a reconstructed image, intertwining certain real and imaginary experiences that are flowing in their minds.

Acting global, detaching local?

While the 'high-tech astronaut families' engage in their transnational network and enjoy the reconstructed 'Taiwan' image in their minds, their place-related community ties seem to fade away. In Silicon Valley, the Taiwanese 'high-tech families' – a marginalized group in America – prefer to disperse into American suburbs and become Americans, so the Chinatown environment becomes alien territory to them. None of the interviewees cared to become neighbours with Taiwanese people. Some of them even mentioned that they were afraid of living in a neighbourhood dominated by Asians, especially Chinese. Predominantly white, suburban neighbourhoods came to embody 'a sense of home' to them for two reasons. First, a suburban home provided socio-spatial security and, second, the suburban habitat conveyed mainstream American cultural values much more openly than residence in the inner city.

Their high-tech occupations notwithstanding, what happened in Silicon Valley is very similar to the events experienced at Monterey Park.[12] In the 1980s Monterey Park was established outside the city limits of Los Angeles as the first suburban Chinatown (Fong 1994; Horton 1995). From Monterey Park in the Los Angeles basin to Cupertino and Fremont in Silicon Valley, the confrontation between old-timers and newcomers always started with the issue of promoting bilingual language signs. Based on my survey, the groups employed in the high-tech sector in California produced suburban commercial enclaves that were spatially the same as other suburban malls, but with Chinese characters and offering almost exclusively Chinese items and services. In addition, there were typical upper-market houses laid out according to *feng shui* principles, with lucky door numbers, symbols and often gardens cultivated in Asian styles. These unfamiliar features caused racial and class tensions, but had minimal physical impact on the overall suburban development pattern in California. More importantly, the Taiwanese 'high-tech families' are unaware of the multicultural conflicts created by their behaviour. They criticize Chinatown and the Chinese immigrants sometimes even more harshly than they do other ethnic groups. The following is an example from my field notes:

> Living in a Chinese-dominated neighbourhood, Henry feels threatened. He is not concerned with the school district since he is still single. Nor does he consider the ethnic diversity in his home-buying list. He feels threatened if the neighbourhood has many Chinese residents. His remark, 'They are noisy because they like to speak loudly. Then, it is dangerous to be a neighbour with them because they do not follow the traffic rules when they drive. You know, if you hear your Mexican neighbour finally buys car insurance, it means there must be some Chinese moving into your neighbourhood.' His jokes about this were malicious and yet his sentiments did not seem entirely false. He also cannot accept that Chinese neighbours like to occupy parking spaces and are seldom seen to pick up their litter.

As with other socially marginalized groups, the Taiwanese employed in the high-tech sector feel strongly that a favourable location for the suburban home moves their social position from the edge to the centre. Significantly, good school districts are mostly associated with these communities. Good school districts provide high-quality education that promises a prosperous future for their children. However, some interviewees pointed out that Chinese Americans also tended to move to good school districts and new development communities. Thus, sooner or later, they create Asian-dominated blocs. One of my interviewees explained that she has lived in the suburban community of Cupertino for four years. During this period, many long-term Caucasian residents had moved out as a result of many Asian newcomers having moved in. Hence, former residents of Hong Kong, Taiwanese and Asian Indians now form the majority. The eleven families in her block consist of three Indian, two Taiwanese, three Caucasian and three former Hong Kong resident households. However, they seldom interact with each other.

In terms of the Hsinchu Science Park, at the press launch of *The Sorrow of the High-tech Colony* held in April 2000, the mayor of Hsinchu explained how the local residents have been affected by the socio-environmental impact of rapid high-tech industrial development. Local governments can no longer deal with the impact of heavy traffic, the crisis of rapidly overcrowding school districts, toxic water pollution and public health crises, to the point where local government and local communities desperately seek help from global high-tech cooperatives. The local people request that high-tech companies improve public facilities and review their environmental strategies. However, due to political tension between the city government and the Hsinchu Science Park Administration, their petitions usually turn into legal battles between grassroots activists, local government and the high-tech companies (Chang *et al.* 2001).

Many local Hsinchu citizens also work inside the Park as manufacturing workers. As previously discussed, the unprecedented success of the high-tech industrial developments in the Hsinchu Science Park created 90,000 jobs in a decade. It has drawn approximately 15,000 'high-tech families' as newcomers; the remaining workers are labourers from the Hsinchu region. Although the Hsinchu Science Park benefits most local residents economically, the indigenous population and local 'high-tech families' feel divided by the wall.[13] One community designer who is also a grassroots activist states:

> There is a love/hate relationship between the Park and us [local residents]. Many local Hsinchu residents actually work inside the Park.[14] I disagree with many of the attitudes the Hsinchu Science Park Administration presents to our Hsinchu City. However, my younger brother and my sister-in-law both work in a high-tech company there [in the Hsinchu Science Park]. They get great salaries and benefits from their companies. The Congresswoman you just interviewed has strongly protested the water pollution issue that is created by the Park. You know, her husband works for the Park, too. He is a senior R&D staff member with a very high status in the Industrial Technology Research Institute. They also bought their house from the Windbond Sweet Home project.

In stark contrast to their detachment from the respective Hsinchu and Silicon Valley local communities, the CEOs of the Taiwan high-tech companies are engaged in both national and international policy-making. They play a key role in shaping high-tech and China–Taiwan policies. Morris Chang (CEO of TSMC) and Robert Tsao (CEO of UMC) were members of the Presidential Advisory Committee for President Chen in 2000. They and other CEOs of high-tech companies are also opinion-leaders in Taiwan civil society. Moreover, they donate a great deal of money to support the country's performance at an international level, nationwide religious activities and the information technology related to expanding plans for high-ranking universities in Taiwan (Chang *et al.* 2001: 44).

Conclusion

My research began by asking where, in the context of a transnational information age, 'high-tech Taiwanese astronaut families' identified as being their home. The investigation of the relationship between home identities, lifestyles and landscapes demonstrates explicitly how economic transformations have an impact on cultural transformations and all consequent environmental transformations. With the analyses of transnational life split between Silicon Valley and the Hsinchu Science Park, I concluded that the socio-spatial structures of the two places are becoming increasingly similar and interconnected, while diverging from and relating less to their local contexts. In the context of transnational migration, immigrants commute between their new and old 'homelands', maintaining two homes and double lives, seamlessly and simultaneously co-existing across two societies and two continents. Inevitably, the residential experience in one place blends with the other. The specific phenomenon is apparent: Transnational social networks and lifestyles establish a separate sense of transnational community that is separate from local communities.

Limited by the time and scope of the study, my current research solely investigates the phenomenon initiated by the Taiwanese and taking place in the high-tech industrial centres of the Hsinchu Science Park and Silicon Valley. It is essential to conduct longitudinal studies to understand the long-term effects of home-identity transformations initiated by high-tech industrial development in this era of globalization. The time-frame should encompass at least one and preferably two decades. The spatial scale should be across the Pacific Rim, collating information from high-tech centres in China, Singapore, Malaysia and Japan, and also from India, along with information from the trans-Atlantic countries of France, Germany, Portugal and Spain.

The physical development of high-tech industrial centres brings not only economic and technological transformations to a society, but also the juxtaposition of social identities and cultural values. Social exclusion and inequality can be expected to continue under these conditions. More alarming is the sprawling Silicon Valley look-alike landscapes that are burgeoning internationally. Policy-makers, planners and designers must not be naive, nor can they take a short-term view of the generic high-tech centres proliferating around the globe.

Notes

1 A pioneering Taiwanese planner who specializes in industrial parks prepared the project. She explains:

> The Chair of Hsinchu Science Park Administration invited me to join the Park planning, because I was the only Taiwanese who had professional experiences at planning and designing high-tech industrial parks at that time. I used to work in a planning firm in Palo Alto with a specialty at developing industrial parks. I moved back from Palo Alto with my husband who is a high-tech engineer. He got a pretty good job in the Park, so we decided to move back. The Park Chair invited me to give a lecture regarding high-tech park planning and develop-

ment in Silicon Valley, since no one knew how to develop a science park in Taiwan. Then, they asked me if I was interested in working with them.
2 Before the Hakanese migrated to this area, Chin-shan-mian was inhabited by an indigenous population. Many confrontations between the two groups occurred during the Hakanese migration period, but this cannot be covered in this article. For detailed information, consult Lee *et al.* (1997) and Wu (1998).
3 According to the Chin-shan-mian community leader, Mr Wu, the Hsinchu Science Park Administration moved no less than six community gods in the relocation of one new temple. Historically, different local villages had worshipped their own community god. Through the relocation, historical religious networks were interrupted and local residents separated from their home villages.
4 According to Lee *et al.* (1997: 77), every irrigation pond has a name and is supported by a nearby neighbourhood. All of the irrigation ponds have their own legends, but since only senior local residents remember these stories, this environmental history will soon fade away. In the meantime, high-tech industrialization has transformed the farm land and agricultural population: Chin-shan-mian is now a high-tech sector.
5 From a telephone conversation on 27 February 1998, repeated in more detail on 21 March 1998.
6 According to the club's guidelines, employees and residents in the Hsinchu Science Park are eligible for membership of the housewives club. People outside the Hsinchu Science Park circle have to be introduced by three or more members of the club to gain membership. Their applications also have to be approved by the board members. The founding member and some other informants have realized that this rule creates a segregation between the actual local community residents and the Park community. However, they do not intend to change this rule. Instead, they declare: 'Our members should participate in the local communities more.'
7 'Green card moms' refers to those wives who have applied for a green card from the USA and are waiting to become US citizens.
8 Equal to around US$180–US$250.
9 I was unable to go, because it was not within my budget. However, the Alaska cruise package was a very popular one in the summer of 1998. My interviewee told me: 'It's only US$1,500 per person. It's a great discount price that you can get only from Taiwanese travel agents.' Another interviewee e-mailed me, asking for information regarding summer high-school programmes at UC Berkeley.
10 I participated in their group chats. At one gathering, they introduced me as the woman who interviewed their friends at Bamboo Village in Hsinchu Science Park. Then they introduced a member of the group to me as the one who had successfully lost weight by undergoing acupuncture treatment. During their chat, one housewife spoke of a person who lived in a certain district of Cupertino, enabling her children to enrol at a certain school. Another housewife insisted that this could not be so, because she knew that the school district in question did not cover that area. Both of them insisted that they were correct. They then made a telephone call to clarify the situation, but the person they called was not at home.
11 I performed three informal experiments to see how the Taiwanese and the Americans respond to branches of *Ranch 99*. I explained to Americans how *Ranch 99* symbolizes 'a sense of home' to Taiwanese immigrants. The typical response I received from them was: 'Oh! I never realized *Ranch 99* was so different, even though I pass it frequently.' I also brought three types of Taiwanese tourists to *Ranch 99*. The first group comprised people who had just arrived in San Francisco. The second group had been in America for a while, but did not live in the Bay area. The third group had been travelling in California for about a month. All of them had the same response: 'This is nothing but Taiwan!'

12 Monterey Park used to be a city with a predominantly white and Hispanic population. Today, more than half of the population is Asian, mainly Chinese immigrants from Taiwan, Hong Kong and China. The 'Monterey Park syndrome' refers to the way Chinese newcomers took over the good school districts of the suburban communities in Monterey Park city in the early 1980s. Chinese commerce, following the immigrant Chinese population, squeezed into the traditional American business districts and gradually dominated the area. The old-timers felt threatened by foreign commercial signs and there were other multicultural conflicts over such daily issues as residential density, attitudes towards education and the preservation of the historical community.

13 My field notes show that the Hsinchu Science Park employees who have junior college or high-school degrees are from the local Hsinchu region. These employees are high-tech manufacturing workers employed on assembly lines. According to one source (http://sipa.gov.tw/en/seconde/hsip/hsi11000_08.htm, downloaded 8 April 2000), on average 55 per cent of the workers annually are local (Hsinchu Science Park 1999).

14 He estimates that about 50,000 Hsinchu Science Park labourers are Hsinchu residents. According to the annual population report, the total population in Hsinchu City was 379,000 in 1997. The total population in 1996 was 345,000, with 120,000 citizens between the ages of twenty and thirty-five (*Taiwan-Fukien Demography Quarterly* 1997).

References

Anderson, Kay J. (1987) 'The idea of Chinatown: the power and institutional practice in the making of a racial category', *Annals of Association of American Geographers* 77(4): 580–98.

—— (1991) *Vancouver's Chinatown: Racial discourse in Canada, 1895–1980* (McGill-Queen's Studies in Ethnic History, ed. D.H. Akenson), London: McGill-Queen's University Press.

Chang, Shenglin (2000) 'Real life at virtual home: silicon landscape construction in response to the transcultural home identities', unpublished Ph.D. thesis, University of California, Berkeley.

Chang, Shenglin, Tu, Wen-ling, Yang, Wen-chuan and Yang, Li-fang (2001) *A Study of the Environmental and Social Aspects of Taiwanese and US Companies in the Hsinchu Science-based Industrial Park*, report for the California Global Corporate Accountability Project, commissioned by Nautilus Institute for Security and Sustainable Development, the Natural Heritage Institute and Human Rights Advocates, Berkeley.

Fong, Timothy (1994) *The First Suburban Chinatown: The remaking of Monterey Park, California*, Philadelphia: Temple University Press.

Horton, John (1995) *The Politics of Diversity: Immigration, resistance, and change in Monterey Park, California*, Philadelphia: Temple University Press.

Hsinchu Science Park (1999) 'Employees by education and their average age', Hsinchu Science-based Industrial Park Administration, www.sipa.gov.tw/en/seconde/hsip/hsi11000_08.htm (downloaded 8 April 2000).

—— (2001) Hsinchu Science-based Industrial Park Administration, www.sipa.gov.tw/en/seconde/hsip/hsi19981_01_06.htm (downloaded 1 September 2001).

Lee, Ding-zan *et al.* (1997) *Community History of Chin-shan-mian*, Hsinchu City: Hsinchu City Culture Center.

Mozingo, Louise (2002) 'Where "talented young men are encouraged to think freely": inventing the corporate campus, 1930–1965', unpublished manuscript.

Soja, Edward W. (1996) *Third Place: Journals to Los Angeles and other real-and-imagined places*, Oxford: Blackwell.

Tseng, Shiao-yah (1995) 'The aesthetics of daily life and community development', unpublished M.A. thesis, Ching-Hwa University, Hsinchu, Taiwan.

Tuan, Yi-Fu (1996) *Cosmos and Hearth: A cosmopolite's viewpoint*, Minneapolis: University of Minnesota Press.

Welsch, Wolfgang (1999) 'Transculturality: the puzzling form of culture today', in Mike Featherstone and Scott Lash (eds), *Spaces of Culture*, London: Sage.

Wu, Ching-jie (1998) *The Orchid in Silicon Valley: Chin-shan-mian*, Hsinchu City: Jin-shan-mian Culture and History Research Studio and the East Hsinchu Jin-shan-mian Community Development Association.

10 To cross or not to cross the boundaries in a small multi-ethnic area of the city of Tehran

Soheila Shahshahani

Introduction

Various authors give different numbers of ethnic groups in Iran. None of the authors include every group. The reason for particular inclusions and exclusions is the availability of material in a European language, not the existence or non-existence of populations. Here I will mention two such documents. The first is by Orywal (1986), which mentions forty-one different ethnic groups. There is only one entry for the Kurds, although there are four distinct Kurdish dialects in Iran. The Komanchi group is mentioned because research exists on this very small pastoral nomadic group. If there is reference to this group, then the Fars or Persian-speaking language population should be divided into dozens of groups across the various provinces of Khorassan, Golestan, Tehran, Hamadan, Esfahan, Shiraz, Yazd, Kerman and Bandar Abas. A second source is the *Atlas d'Iran* (Hourcade *et al.* 1998: 61), which mentions thirty-four ethnic groups in Iran.

Which of these ethnic groups can be found in the small area of Tehran that I have set out to study (Map 10.1)? The original inhabitants were Fars (Persian speaking), either from the region itself or from the city of Kashan and its vicinity in central Iran. Today, among the 'temporary' residents, there are Kurds, Turks and Afghans (from Hezarah). Street names have become ways of commemorating in Iran. So through a study of street names in this area we can gather certain indications about their inhabitants. We shall see this in some detail later. An exception is a cluster of three street names originating from southeastern Iran. Since no one recalls any inhabitants from this region, this is a matter to ponder on. Sunnit Kurds had also lived here many years ago. However, what is very distinct in this area of northern Tehran and typical of the region is the division between three different populations (explained later). Northern Tehran was a cluster of little villages with rural inhabitants until the middle of this century and that population has continued its existence in the area; some villages also had big landowners who lived in mansions. The native population continues to occupy this area, although now possessing less property. Finally there has been an influx of new residents. Some have bought or built houses and live permanently in this area. Some have rented a flat in one of the newly built apartments. These constitute the 'temporary' population.

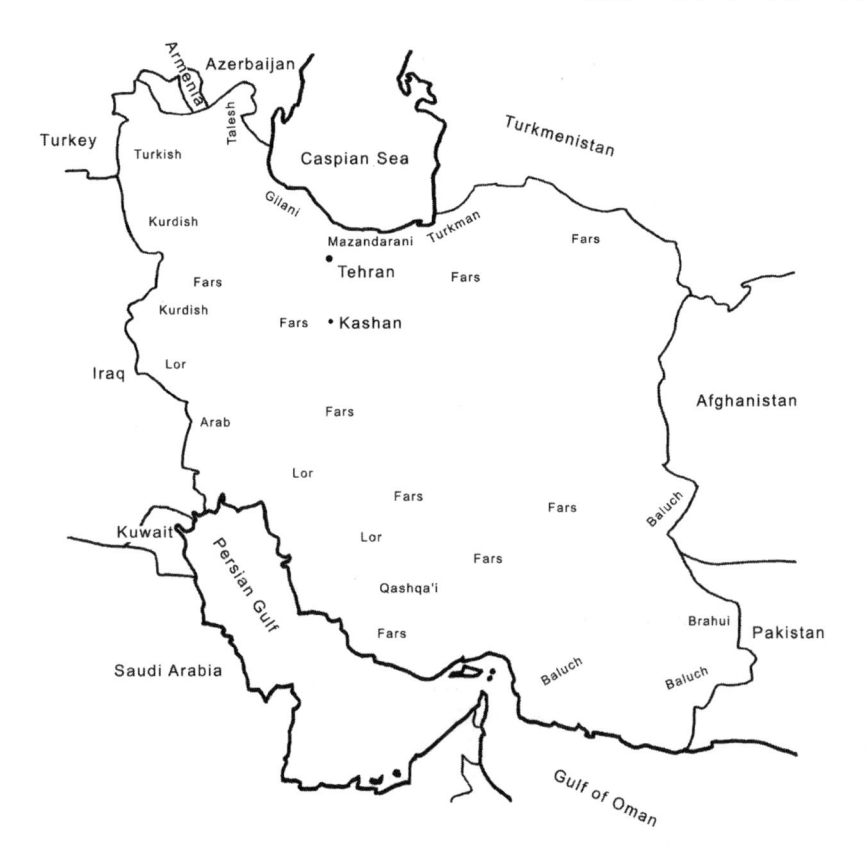

Map 10.1 Ethnic groups in Iran

There is an ongoing relationship between the people who have come from different regions of Iran and their original location. There is also a relationship with those outside Iran. This is not only between the Afghans and their homeland, but also between residents and members of their family who have left the country to live in Europe or the USA. Thus, in this very small area, we find different styles of living: migrant, rural, urban middle-class and cosmopolitan urbanites – that is, those who maintain contacts and similar lifestyles through the internet and other mass media (Castells 1989). Ethnic identities overlap with class differences.

In this chapter we shall see how these different groups communicate, what their boundaries are, where and when they cross these boundaries. Inner identity is defined in relation to others. It is dialogical: 'people manage to shape dialogical identities while at the same time reify monological ones' (Baumann 1999: 40). This is not a process that takes place in isolation, rather it is 'negotiate[d ...] through dialogue, partly overt, partly internal, with others' (Taylor 1994: 34). In this study of a small area of the mega-city of Tehran, we can see the dynamics of an old culture that has always boasted of the peaceful cohabitation of its various ethnic, class and religious affiliations. Theoretically speaking,

> Instead of viewing society as a patchwork of five or fifty cultural groups, it [processual discourse] views social life as an elastic and criss-crossing web of multiple identification.
>
> (Baumann 1999: 138)

This is how we can come to an understanding of daily life in this northern area of Tehran.

The history of Ekhtiarieh

First we shall study the area, called Ekhtiarieh, look at its history, its occupants and their ethnic/class identity. Ekhtiarieh is located in northern Tehran, in the first and third sections of an urban division. Ekhtiarieh Street is divided into North and South, with a square marking the division (Map 10.2). The crookedness of the streets recalls rural alleys. Once in a while, one comes across a cluster of tall poplar trees, confirming again its rural pedigree. Ekhtiarieh, like many areas of northern Tehran, was formerly a village that was totally or partially owned by an aristocrat resident of Tehran. This arrangement provided the family with a summer home. A rural population lived in the area all the year round.

The name 'Ekhtiarieh' comes from the title of the owner of a large part of this village, then known as Rostam Abad. There used to be a street named after him: Saheb Ekhtiar Street; but his actual name was Gholam Hossein Ghaffari. Saheb Ekhtiar (meaning 'possessor of authority') was the grandson of Farokh Khan Ghaffari (Qaffari), entitled Amin od-Dowleh, who was born in 1814 (*Safarname* 1361: 22). The Ghaffari family had an ancient pedigree, going back to Abuzar, a contemporary of the prophet Mohammad.[1]

Farokh Khan had come to Tehran in his early youth and entered the court of Fath Ali Shah, who regarded him as a loyal and intelligent man, endowing him with the title Amin ('the Honest'). Amin od-Dowleh became a special servant to the king and was sent on military or civilian missions to Herat, Esfahan or Fars. He later became the king's treasurer and in 1856 went to Europe as a special envoy.[2] Back in Kashan and Tehran, he commissioned the construction of a number of monumental buildings. He was also instrumental in introducing many modernization measures to Iran, including the telegraph system and the project of sending students abroad to study science, technology and the arts. He died in 1871 and was buried in the city of Qom in the family mausoleum. Afterwards, Amin od-Dowleh's brother became the minister to the court of Nasser ed-Din Shah. His son was Gholam Hossein Ghaffari, better known as Saheb Ekhtiar. He and his cousins also had political responsibilities at the court. Hassan Ali Mo'aven od-Dowleh, a grandson of Saheb Ekhtiar, became especially prominent at the court.

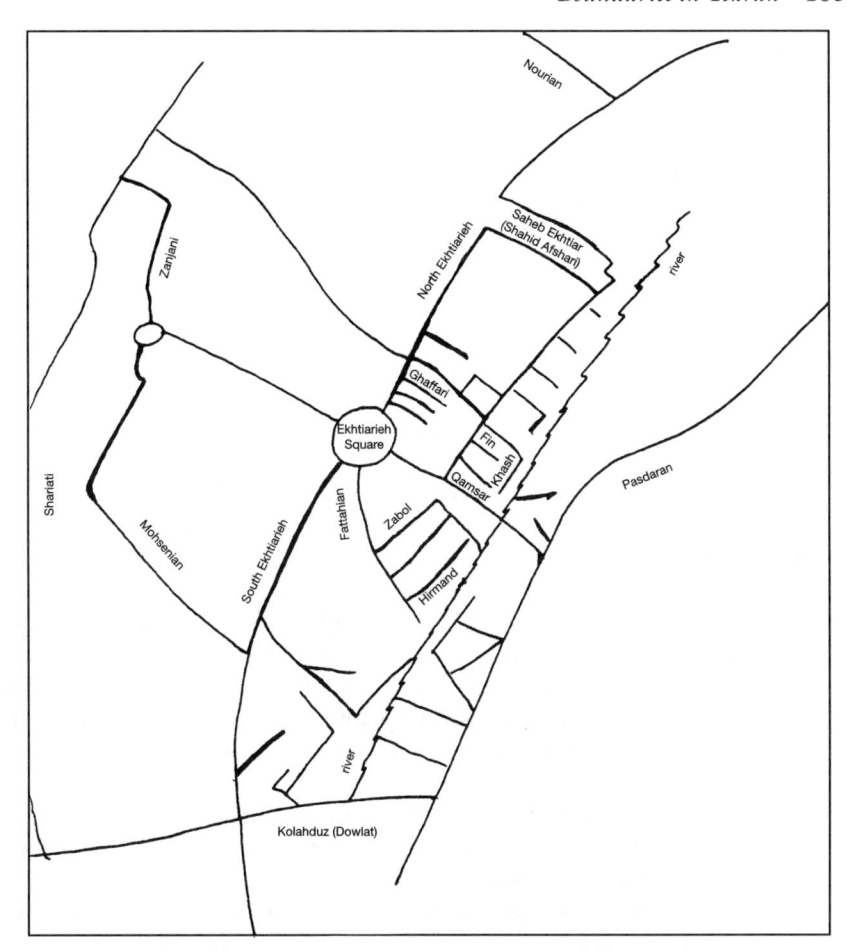

Map 10.2 Ekhtiarieh

The Ghaffari family has inherited many handwritten orders and letters from the king, along with some fine china made in Ghaffari's honour on which his name was engraved. Lots of documents from his office have been used in editing books, though some were confiscated after the Revolution. A direct descendant of the Ghaffari family still lives on Saheb Ekhtiar Street, which is today called Shahid Afshari Boulevard, having been renamed after a martyr of the Iran–Iraq War.

History and material culture separates the Ghaffari family from other groups in the area. Their pedigree is documented and is further vouched for by plates and other objects that bear their names. The identity given to this family, through these objects and their shared memory, is a means of identifying and delineating them. Their roots in the area and in Kashan do not make them a conservative group: Their modernist stance has been attested by Farokh Khan

Ghaffari, who was instrumental in the mid-nineteenth century modernization of Iran. Today, the family still maintains contact with its members living abroad.

What other families lived in Ekhtiarieh? One was the Mohtasham family, who provided secretarial services to the Saheb Ekhtiar family. They still live in Ekhtiarieh, but the street that bore their name was renamed after the Revolution. Another family, the Moqqadam – a family related to the court – had a piece of property in the area. Before the Revolution two other families – who also had aristocratic origins – came to live here. Finally, an intellectual family with an important pedigree came to reside in Ekhtiarieh: Ayatollah Sheikh Fazlolah Nouri, an important figure of the Constitutional Period, was a grandfather of this family. The families who came to live in Ekhtiarieh after the Revolution cannot boast of an important pedigree, being *nouveaux riches*.

Who were the other residents of the area? There were five families – the Rahimian, the Zanjani, the Nourian, the Mohsenian and the Fattahian families – who were all small landowners. They still live in the area; all of them have streets named after them. The majority of the population, however, were peasant-herders and workers. They inhabited a smaller area called Shoqal Abad, which is located to the east of the Saheb Ekhtiar property, on a slope towards a river. People recall, related to this area, farms of wheat and barley, vineyards, orchards of walnuts, pomegranates and so on. The vineyards attracted jackals from the neighbouring mountains, which explains the name (*shoqal* means 'jackal'). Today, 'Shoqal Abad' is used pejoratively to mark a class distinction. The houses on the slope towards the river and in the alleys are small – about 100 square metres each.

The current occupants of the area are descendants either of these former local residents or of those who were brought to the area as gardeners or servants to the Ghaffari family from the vicinity of the city of Kashan, the area of their origin. That is why we still find a strong group of related families who have come from the region near Kashan – that is, from Qamsar, Fin, Meymeh or the village of Kamu. Others have come from Varameen, where the Ghaffari family also had some property. This population calls itself the 'stables' (*sabet*) as opposed to the 'temporary', those who have come to occupy the area since 1947, when the land was subdivided into small parcels and sold to incoming residents. This dichotomy in terminology ('stables' and 'temporary') is the Shoqal Abadi way of distinguishing among themselves and separating their group from the new middle class who have come more recently to occupy their area.

Some residents recall having come here for holidays, on weekend trips or for a picnic on *sizda bedar*, the New Year holiday. On such occasions, according to local people, 'hundreds of families would be picnicking around the pool of water, under enormous trees'. Weddings that took place here are still the most memorable events. The Ghaffaris had carriages for commuting to Tehran, but since 1947 there has been a bus service, making the area more easily accessible. The rural residents of the area were relatively self-sufficient, cultivating their own food, and got other necessities through the few local stores. There was a bath, a teahouse and a cemetery in the vicinity.

Another important fact about the area is the presence of a military base and an ammunitions factory, owned by the Ministry of Defence. The children of the old residents of the area and many of the 'temporary' people used to work at the factory. Two different categories of military population lived here. One consisted of officers of the upper rank, some of whom were married to German women. The others were low-ranking or non-commissioned officers. The former category had the air of the closed upper-middle class, with a foreign tone. There are still people who recall a visit by Pahlavi II and his first wife, Fowzieh, to the military base. 'I remember she did not have any stockings!' says an elderly woman, thus emphasizing the difference between the religious and non-religious tendencies of the people, not only in this area but over the whole country.

Low-income housing had been developed for the lower-ranking military personnel. The military workers were under constant surveillance and no form of political activity was allowed to develop. During the past ten years, with the 'tower building fever' in Tehran – that is, the construction of residential high-rise buildings in the northern part of the city – some of these old houses have been replaced with high-rise buildings for the new middle or upper-middle classes. This has changed the rural aspect of a few streets.

Finally, there is also a cluster of Afghan workers in the area. One group lives on the boulevard, in an apartment which has been under construction since 1991 and is not yet completed. With this group we now have four different classes formed by different ethnic groups living in Ekhtiarieh.

The area is totally Shiite, though the residents of Shoqal Abad recall a Sunni Kurd family who used to live on their street. They lived there for fifteen years and it was only with regard to them that the inhabitants discussed any religious and language differences in the area. Even the present Afghan residents are Shiites.

In fact, the area is a tribute to the names of several people, made possible by Saheb Ekhtiar. He had bought this area, Ekhtiarieh, in the territory of the village of Rostam Abad and had three stone buildings constructed. Besides tending to the irrigation, which was done through the *qanat* system (Shahshahani 1998: 24), he provided other facilities. He resided in Ekhtiarieh and registered his property in 1925 at the General Registration of Properties. Later he and his descendants donated land to the city for the development of streets, the square and the boulevard bordering their property. As an influential person he helped the people of the area to further their careers. Although history, material culture and common memory are means of defining the identity of a group, crossing the identifying borders had been necessary in order to keep in contact with the other occupants of the area. It was through this contact, made in specific defined ways, places and times, that the Ghaffari family could become aware of the needs of the people of the area and felt a responsibility to respond to them.

So far, we have the history of the area and the residents. What about their daily life? This information was gathered through questionnaires that I filled out myself, supplemented with my daily observations and another special questionnaire for particular people who contributed to the history of the area presented above.

The daily life of the residents

The Afghans

The population of the Afghan group fluctuates every few months. In the winter of 2000 there were seven men, all of whom were related. They lived in two rooms (30 square metres), allowing for about 4.2 square metres for every individual. They have other relatives in various cities of Iran. The duration of each man's stay in Iran ranges from three to nineteen years. They work in various jobs, all related to construction. Each of them spends about 22,000T (tomans) per month on food and clothing, trying to save 30,000–35,000T a month which they send to their families in Pakistan or Afghanistan. Some families residing on the boulevard use the services of the Afghani group for handywork, cleaning and so on. They are considered trustworthy and cheaper than Iranian maintenance workers. They mix with the youth of the area who join them to play football on the boulevard in the evenings. One of the Afghani men is very athletic; he is admired by the youngsters when he does his exercises at night. This is a very good opportunity for socializing. Until late at night, youngsters can be on the central lawn, doing exercises and chatting. During the month of Moharram[3] and for certain religious ceremonies, some of the Afghans might go to the local mosque or are invited to specific houses. They are often visited by other compatriots residing in Tehran or in other cities of Iran, if not abroad.

They do their shopping in the area or at a popular section of Tehran (the Enqelab Square) and one of them frequently receives parcels from a brother in Germany. For them, shopping is an interesting way to pass the time. Having to purchase material and utensils for building and repair puts them in contact with the local shops. The most prized objects in their home are the refrigerator, colour television and video player. They have nothing from Afghanistan except finely embroidered undershirts made by their female kin. They have a telephone, which they use quite a lot to call their family members in Pakistan or in Germany. They have travelled to various cities of Iran – such as Mashhad, Shiraz, Esfahan and Bandar Abas – either to work or on pilgrimage. As most of them have physical problems – injuries from the Afghan conflicts or from accidents at work – they go for special medical treatment, something that can be very costly. They have dreams about going to work or acquiring the right of abode in a Western country. As many of their compatriots have gone through Turkey to Europe, this is not a very far-fetched idea. At least it is a dream and exciting idea that they can speak about. The experience of those who have gone and those who have not been successful and returned, make for many hours of talking, thinking, planning and dreaming. Meanwhile, they go almost every year to Pakistan or Afghanistan to visit their families. They are the 'deterritorialized' (Appadurai 1991: 192), who share with our fourth group, the cosmopolitan urbanites (see *The inhabitants of northern Ekhtiarieh* below), certain aspects of the 'informational city'.

The inhabitants of Shoqal Abad

Our second group consists of the inhabitants of Shoqal Abad. Small houses with yards, in which two related families live on different floors, are prevalent. Here, between two and six people live in an area of 55 to 120 square metres; that is about 28 square metres for every person or 95 square metres for each family. The minimum income they consider necessary to live ranges from 30,000T to 150,000T per month, which is about 40,000T per person and 110,000T per family. They say that they would need less for themselves, but since they often have visitors, they have to spend this much.

As mentioned above, they come from the vicinity of Kashan or from villages close to Tehran such as Dar Abad or Ozgol. Those who come from the Kashan region still have contacts there; they receive foodstuffs, such as dried or seasonal fruit and some dairy products. Some have carpets from Kashan. Their most prized objects are these very carpets[4] and the trio of refrigerator, television and video player.

As they go every morning or evening to the shopping centre at Ekhtiarieh Square, they have the opportunity to meet neighbours. When and if men do the shopping, which is rare, then women leave the house less often and have less occasion to meet others. Most men and women go to the mosque for the evening prayer, where praying and socializing is sexually segregated. When an announcement is made at the mosque or a flag is seen in the vicinity, women go to listen. Certain houses are known for their annual activities and offerings. Visiting houses is restricted to close relatives; neighbours rarely visit each other. There is hardly an isolated family in this group. They all have relatives in the same house or have at least two relatives living in the area. So daily visits to relatives are common. Children play in the alleys, though a favourite place is the boulevard with its central lawn where they can mix with the children who live there. The latter group does not welcome this situation, but since the number of the 'trespassers' is restricted, they accept it. But the social distance between the two groups is strictly observed, with no house visits issuing from this mixing on the boulevard or at school.

Besides regular trips to their native region of Kashan, a pilgrimage to Mashhad is the most frequent voyage undertaken. Thus, depending on personal means, only a few cities – such as Esfahan, Bandar Abas, Ahvaz and Abadan – the Caspian Coast and Tabriz are visited. Only the two richer members of this group – one an unmarried woman and her mother, the other a man who has a car radiator repair shop – have gone further afield: to Mecca, Kuwait, Syria and Turkey.

The 'temporary' group in Ekhtiarieh

Now we come to the 'temporary' group – those who are in Ekhtiarieh for a number of years but will leave sooner or later. Their choice of this location is due to special circumstances. Every family (three to five members) lives in an area of 60 to 400 square metres, with about 45 square metres for every person. They need at least 120,000 to 600,000T per family per month, that is about 80,000T per person. If they have grown-up children, extra-curricular tuition for the children is mentioned as an important item of expenditure.

These families do most of their shopping at places accessible by car; in this case, the women may be accompanied by their husbands who do the driving. It is also possible that the husband does the shopping alone. None of these families go to the local mosque, but they do go to religious ceremonies at specific houses, often on invitation, outside the area. The offerings they receive at such ceremonies may encourage them to go. Their children go to local schools and mix with others, at times going to their houses. They feel isolated in the area as they are individual families; this gives them reason to socialize with their neighbours. Although they may have very different social standing or belief systems, loneliness pushes them into socialization with the other residents of their apartment complex. These relationships often break up after moving from the area.

Those who come from different cities of Iran bring along their local carpets; besides this, gold – in the form of coins and jewellery – is mentioned as another prized object in the household. Crystal vases and vessels have been an important acquisition ever since the end of the Iran–Iraq War; in one household, there was an overflow of it, all over the place. Artificial flowers (called 'Chinese flowers') are also a new addition to the decoration. Plastic flowers are the norm in Shoqal Abad households, whereas fresh flowers can be seen in the houses on the boulevard.

Visiting or visits from relatives living elsewhere in the city occurs at least once a week. The telephone is used much more frequently in these households. It is used not only for business, but also to get family news from those who live far away. Family visits from the provinces are also a common occurrence, particularly during the New Year holidays and the summer vacation. Their travel itineraries are more varied than the previous group; they are less pilgrimage-oriented and more touristic. The memory of a trip out of Iran – for work or as a tourist – will be recollected with excitement.

The inhabitants of northern Ekhtiarieh

The fourth group is the inhabitants of northern Ekhtiarieh, living around the boulevard. Each family lives in a dwelling of, on the average, 277 square metres, with 93 square metres per person. As far as their expenditure is concerned, they need about 500,000T per month for a family of four. Many earn much more than they spend. The women of these households do not have a precise idea of the income of their husbands; this was an exceptional case in Shoqal Abad. Here, the most prized objects have an emotional as well as monetary value attached to them. Antiques, paintings, pianos and family relics are mentioned as the most important objects, besides jewellery, silver vessels and carpets.

These households generally prefer apartment housing, with at least two kin living in other flats. Thus there is the same kin relationship observed in Shoqal Abad. Other relatives do not live in the area but are scattered around the city or in the provinces. Among this group, visits from friends were frequently mentioned. A few of the families hold religious ceremonies at their homes. They never go to the local mosque, although they may have helped financially when it was being built and repaired. If they go to religious ceremonies, it is on invitation or as an annual ritual. Celebrations of marriage, birth and particularly condolences during mourning are common occasions for lots of visits. As these families have many connections in the city, the death (and mourning) of a family member is announced publicly; no specific invitation is required. Since it is considered an obligation to sympathize with those in such distress, mourning is a frequent reason for visits, becoming an event for friends and family to meet. It is an occasion when class and ethnic boundaries are crossed.

I observed two extreme forms of passing the time for elderly women. One was gambling, which occupied one elderly person a few times a week. The other one was engaged in volunteer work for children in need, such as runaways and those in correctional houses, or simply for those who were illiterate. The schools for children are not necessarily in the neighbourhood. As parents want good education for their children, they are willing to enrol them in distant schools. They also send or take their children to various extra-curricular activities, such as physical education and courses in language, the arts and so on. The mothers often use the family car or their own car for such chores. Women generally drive cars and they take care of the family shopping. This is considered a chore, unless it is window shopping with friends for clothing and house utensils. Something which has become popular since the Revolution is going to fashion shows (dress and jewellery) or garage sales. Hence, shopping as a daily necessity is not an occasion to socialize with neighbours as it was with the second group. Members of this group take the car and drive to various stores that are considered particularly good, which may even take them totally outside Ekhtiarieh.

All members of this fourth category have relatives that live outside Iran and their telephone bills vouch for such contacts. They have travelled far and wide in Iran and abroad to various countries in the West, with perhaps a few visits to the Far East. Some have studied abroad. Some are connected through e-mail; one person, a university professor, was instrumental in establishing the internet service in Iran. There is also an important publisher among them. Some work with foreign firms. They have an international outlook and their imagination corresponds closely with reality as they have constant global relations. Hence, we find some cosmopolitan urbanites here in Ekhtiarieh.

With regard to language, the members of Shoqal Abad originating from the rural areas of Kashan speak their dialect when together. For example, a woman who recently married a Shoqal Abadi speaks with her relatives in the Kamu dialect, which is incomprehensible to Farsi speakers. The Afghans also speak their dialect. Turkish and Kurdish are spoken by members of the 'temporary' group. English and French are the second languages of many people in the fourth group; there is the isolated example of an elderly German woman married to a man from this group.

I mentioned before that the people inhabiting the area were all Shiite. I should also mention that there is an Ahl-e Haq family – an esoteric sect centred on Kurdistan, but having followers in many parts of Iran – in the area, which attracts many people to its gatherings. Its activities have taken place with varying degree of openness to the public at different periods. Finally, there is one person from the fourth group who is a follower of SiBaba, an Indian guru, and participates in weekly rituals.

Disputes in the area are interfamilial. There are hardly any difficulties with neighbours that would be a matter of concern. At most, there is avoidance. One case of conflict that was reported to me was between the servant of one household and an Afghani; the former was accusing the young Afghan man of taking his job. This was a very brief disruption in the otherwise harmonious life of the Ekhtiarieh neighbourhood.

Conclusion

Thus we have a very diverse group of people residing in an area consisting of a few blocks. Where, how and when they meet is very closely defined. As far as their everyday life is concerned, they have different incomes, different occupations and differently sized properties. There are also differences in their travel itineraries, their shopping habits, the schooling of their children and their social lives. The first, second and fourth groups meet regularly, since the first two groups offer services to the fourth group. The most important point in crossing the boundaries occurs in the kitchen, a space where women share many hours working and chatting. This is not generally shared across the gender divide. Furthermore, when crossing boundaries when there is an invitation for a meal, the presence of a person as a cook or maid/servant becomes more clearly defined. We should also mention private gardens as places where boundaries are crossed: the gardeners are always men and the employers (men or women) speak to them in the garden. Outside defined working hours, no crossing of boundaries occurs.

As far as the men are concerned, they are more distant from the life at Ekhtiarieh today than before: Their professional life takes men of all categories, except the Afghan migrants, outside the area. They have little chance of meeting each other except for formal salutations exchanged in the streets. Crossing the boundaries occurs if there is socialization, at a family level, between the third and fourth groups. This happens occasionally. But between the members of the third ('temporary') group it is more frequent, with neighbours passing many long winter evenings together. The habit of visiting after dinner is common in this group, both among those who have come from the provinces and those originating in Tehran.

The ritual of offering food on the occasion of the death of a family member, or as a votive offering, is also observed among the inhabitants. Fruit grown in private gardens is also exchanged between houses. The offering is made at the doorstep. Taking back the container is an opportunity for crossing the boundaries when one can engage in small talk. Gifts given on returning from trips are most common within a small group of relatives or people with whom one has close contact. Those who have returned from a trip offer these at the 'welcome home' visits by such people.

The second group is the most cohesive. The women meet in the morning when they go shopping. They go to the mosque together and to religious ceremonies. Men meet in the evening at the mosque. There are also house visits, as many households have relatives in the area. A visit from the provinces that they come from provides an occasion for many visits among relatives and neighbours. Economic, social and medical advice and assistance are offered in small networks.

While contact through men, across classes and ethnic groups, has diminished, it persists through children. The boulevard provides a vacant area in which children and young people pass many hours together. Here they play together, across ethnicity and class, but they do not visit each other's houses unless they are from the same group. They only go to a playmate's door to ring the bell and then wait for them to come out to play. Once in a while, through sarcastic remarks, they remind each other of their differences. They do not even call each other on the phone unless their families visit one another. The World Cup, the Asian Football Games and similar events, especially when Iran is in the tournament and particularly when they win a game, is a fantastic occasion for them to call on each other and gather in this central space to share their joy.

There is regularly one night every year when everyone from different groups can come together on the central boulevard: the 'Fire Feast'. It is actually called *charshanbeh suri* ('Wednesday Feast') and is held on the Tuesday night before the last Wednesday of the Iranian year (which begins on the first day of spring: 21 March). After sunset, a fire is lit and everyone jumps over it. Firecrackers are heard all over the place; simpler versions of fireworks are becoming more common. Most of the people come just to watch. The ritual of jumping over the fire while singing '*zardi-e man az to, sorxi-e to az man*', symbolically transfers sickness and worries to the fire and the jumper receives joy, health and warmth for the next year. Thus we see that an Iranian rite provides the setting for people of different groups to share a common experience in public space. Not everyone joins in, but members of all groups are present; even passersby get involved and make it a memorable event. The processions of the month of Moharram also pass through the area. Members of the second group, mostly the young, join this procession. This is another way for them to maintain group cohesiveness and take over public space for a defined period of time. Occasionally, individuals from the third group may also join them, while others may watch the procession or may disregard its existence.

We see that, on a daily and yearly basis, there are many differences that separate these people. There are also many possibilities for them to meet within specific boundaries, physical or ritual. Children meet on a more relaxed and frequent basis, making everyone more disposed to good neighbourly relations. Ritualistic salutations keep people on good terms, while allowing them to keep their distance. Face-to-face relations are being replaced by 'car-to-car' relations (people saluting each other from within their cars while passing by) or 'car-to-person' relations, which are much more distant.

A wedding in 1999 between two different groups shows clearly that boundaries can be crossed. The girl was from the fourth group. The boy was from the second group, but not from Shoqal Abad. His grandfather had been a landowner, but the boy's father had moved into the industrial sector; the young man was fast becoming rich by working with his father. But crossing the class boundaries was exceptionally difficult, as it involved years of rejection from both families until they finally decided to negotiate and permit this marriage. At the wedding ceremony the differences between the families were very prominent. Yet the example shows clearly that these boundaries are somewhat elastic and that they can be crossed at even the deepest level of social existence – that is, marriage.

Thus we confirm Baumann's quotation, given earlier, that 'people manage to shape dialogical identities while at the same time reify monological ones' (Baumann 1999: 40), finding it totally congruent with Iranian culture, with its malleability and ability to negotiate difficult circumstances. This has made the co-existence of different classes and ethnic groups possible. It is just a matter of precise local knowledge of when, where and how.

Notes

1 They provided Kashan, Qom and Natanz – which are small central cities of Iran – with men of political, poetic, theological and artistic capacity. Kamal ol-Molk, father of modern Iranian painting, was an uncle of Farokh Khan.
2 It is interesting to note that on the occasion of his arrival in France, 'Le Rossignol de Teheran', a quadrille for piano composed by Leon Waldteufel, was performed. In 1857 he received the 'Légion d'honneur' in Paris.
3 The first month of the Islamic calendar is the month of Moharram. During this month, Muslims fast from dawn to sunset. The tenth day of the month of Moharram is called Ashura. This day is spent in mourning by Shiite Muslims, since it is the day that the tragic death of Imam Hossein – the grandson of the Prophet Mohammad and the son of Imam Ali – occurred. During this month, but particularly during the first ten days with the climax on Ashura, prayers and mourning in commemoration of that tragic day are common practice. Offering a meal is also commonplace during this period.
4 Iranian people take off their shoes at the door and walk in their socks or barefoot at home, preferably on floors strewn with carpets. Carpets can be considered status symbols. The size, pattern, origin, kind – that is, whether it is wool or silk, as well as the knots per square decimetre – and number of carpets in a dwelling contribute to the prestige of the household.

References

Appadurai, Arjun (1991) 'Global ethnoscapes: notes and queries for a transnational anthropology', in Richard Gabriel Fox (ed.), *Recapturing Anthropology: Working in the present*, Santa Fe: School of American Research.

Baumann, Gerd (1999) *The Multicultural Riddle: Rethinking national, ethnic, and religious identities*, New York: Routledge.

Castells, Manuel (1989) *The Informational City: Information technology, economic restructuring and the urban–regional process*, Oxford: Basil Blackwell.

Hourcade, Bernard, Mazurek, Hubert, Papoli-Yazdi, Mohammad Hosseyn and Taleghani, Mahmoud (1998) *Atlas d'Iran*, Montpellier: RECLUS, Paris: La documentation française.

Orywal, Erwin (ed.) (1986) *Die ethnische Gruppen Afghanistans: Fallstudien zu Gruppenidentität und Intergruppenbeziehungen* (Beihefte zum Tübinger Atlas des vorderen Orients 70), Wiesbaden: Ludwig Reichert.

Safarname (1361/CE 1982) *Safarname-ye Farokh Khan Amin od-Dowleh* (Travel Accounts of Farokh Khan Amin od-Dowleh), Tehran: Asatir Publications.

Shahshahani, Soheila (1998) 'The Iranian city of Kashan: between the desert and the mountain', in Kiyotaka Aoyagi, Peter J.M. Nas and John W. Traphagan (eds), *Toward Sustainable Cities: Readings in the anthropology of urban environments* (Leiden Development Studies 15), Leiden: Institute of Cultural and Social Studies, Leiden University.

Taylor, Charles (1994) *Multiculturalism: Examining the politics of recognition*, ed. and intro. Amy Gutmann, Princeton: Princeton University Press.

Part III
The micro-level analysis of urban ethnic encounters
The streets and the squares of the cities

11 Repackaging difference
The Korean 'theming' of a shopping street in Osaka, Japan

Jeffry T. Hester

Nestled along Osaka Bay about two thirds of the way – moving west-southwest – down the main Japanese island of Honshū, Osaka is Japan's third largest city. Among its 2.6 million inhabitants are some 110,000 resident Koreans. This may seem a small number when compared to minority groups in many other multi-ethnic cities around the world, but it represents the largest non-Japanese ethnic group in any municipality in Japan. Over a third of these resident Koreans – some 38,000 – are concentrated in one of Osaka's twenty-four wards (*ku*) and comprise about a quarter of the ward's population.

This chapter focuses on an area within this ward, which I pseudonymously refer to as 'Ashihama', and in particular on a shopping street within 'Ashihama' that has undergone a substantial transformation in its built environment, through an organized project to represent and market its distinctive Korean element. This transformation, which I refer to as the 'Korea Town Project', is part of the ongoing rich and contested process of place-making occurring in the area. By 'place-making', I refer to activities by which space, geo-symbolic boundaries, history and memory, and the built environment are mobilized in the representation of social groups, in processes of social identity formation and in a complex cultural politics of belonging.

Ashihama is an area where Koreans and Japanese have shared – or not shared – geographical and social space for four generations, going through some quite radical socio-historical changes: Through periods of Japanese colonialism, war mobilization, post-war devastation and recovery, and now an increasingly 'internationalized' Japan, the position of resident Koreans in Japan has shifted. The Korea Town Project offers something of a reflection of these shifts, as well as a glimpse into their limitations.

In addressing the spatial consequences of ethnic diversity as manifested in this little slice of urban Osaka, I discuss the movement to transform the shopping street in Ashihama, the actual changes in the built environment and the dual commercial–civic ideological framework under which this project was carried out. I conclude with a few suggestions about the discourse on 'living together' with difference that has recently emerged in Japan's internal debate on cultural diversity and how it is reflected in the Korea Town Project. First, I would like to briefly sketch the social and historical context of the Korea Town Project.

Ashihama in Osaka's modernization

Until the late 1910s, Ashihama was an agricultural village in the flood-prone lowlands on the urban periphery. The place name is recorded as early as the eighth century in the Chronicles of Japan (Nihon Shoki), Japan's first history, compiled under imperial orders around 720. Something of the village's past is also recorded in several documents possessed by the local shrine and temple dating from the seventeenth and eighteenth centuries. A few local Japanese families claim descent from the early modern Edo-period (1600–1867) village, creating something of a 'native' or 'newcomer' divide (Robertson 1991) by which the 'native' status confers authority by virtue of depth of rootedness in place, a recurring theme in the place-making practices of the area.[1]

A confluence of historical processes resulted in Ashihama developing into a locality accommodating Japan's largest Korean population. Not long after the Japanese state annexed the Korean peninsula across the Sea of Japan in 1910, Osaka's modern industrialization took off, particularly fed by the economic stimulus of World War I. Ashihama, with its cheap land and strategic location on the perimeter of urban Osaka, was transformed during the early stages of this process, rapidly shifting from a village cluster of households surrounded by paddy fields to a mixed residential/small-scale industrial area. Cheap housing for the continuing influx of workers was quickly thrown together, while small-scale, often family-run factories burgeoned. The area was also rapidly growing into a densely packed slum, with some of the largest concentrations of poor housing in the Osaka area. By the early 1930s, the Ashihama area had the largest concentration of rubber factories in Osaka, along with sizeable glass and metalworking sectors (Sugihara 1998).

The labour needs of the industry were fed through migration from Osaka's rural hinterlands, supplemented by active recruitment on the Korean peninsula. The island of Cheju, Korea's largest island (located about ninety kilometres off the southwestern coast of the Korean peninsula), became a favoured site for recruitment for Osaka's factories. Working for wages below those of their Japanese counterparts, Koreans from Cheju were welcomed as labourers by Japanese capitalists. From 1923 a direct ferry service was established between Cheju and Osaka, helping to increase the number of migrants substantially. While migrants from Cheju were a relatively small minority of the overall number of Koreans coming to Japan, they came to form a substantial part of the Osaka Korean population and the majority of Koreans in the Ashihama area (Sugihara 1998: 77, 98). Through the influx of Korean labourers during the Pacific War (most of them repatriated from August 1945), early post-war migration and that of the period of Japan's high-growth economy, Osaka has become the centre of a diasporic Cheju society, with Ashihama at its core. The movement of Cheju Koreans between Osaka and the island continues today.

While the number of commuting workers is rising in Ashihama, it retains its strong small-scale commercial and industrial sector, with both Koreans and Japanese involved as owner/operators of small-scale firms, often housed on the first floor of a two-storey residence. This characteristic of the neighbourhood has important implications for place-making processes undertaken in the area, in so far as householders – male breadwinners – are largely present in the neighbourhood during the day. In short, their most important social affiliations are framed within the community and locality. The stakes involved and energy expended in territorial issues involving community membership and belonging to locality are consequently higher than in areas where breadwinning is largely carried out at a geographical and social distance from residence.[2]

Ashihama as polysemic signifier

To close my sketch of the socio-historical setting of Ashihama, I must mention the status of the place name itself as a trope illustrative of the complexity of place-making in the area. While evocative of strong sentiments of attachment to place, the place name no longer appears on maps and has no single objective, bounded reference. What began as the name of a village, with a venerable provenance in one of Japan's canonical works of myth and history, by the mid-1920s came to be used over a wide area, well beyond the limits of the old village, as a place name and official postal address. In time, the place name also came to signify to outsiders a dirty, overcrowded and dangerous slum, a stigma from which neither Japanese nor Korean residents were exempt. The place name 'Ashihama' also came to be closely associated with its Korean residents, symbolizing a visible, unassimilated presence of ethnic difference in the midst of an otherwise apparently homogeneous Japanese landscape. This latter mapping of 'Ashihama' could be assigned a positive or a negative evaluation, depending on point of view. Finally, when a new address system was instituted in Osaka in 1973, the place name 'Ashihama' was removed from official maps. While this was carried out by administrative fiat, there was apparently little protest from local residents. Rather, it was, at the time, widely welcomed as an upgrading of the image of the place.

While it is no longer in use as an official postal address or administrative district, 'Ashihama' retains its multiple and powerful historical associations for local residents. Following its official erasure, the existing Young Men's Association associated with the local Shinto shrine, a private membership group with its roots in the old Ashihama Village, was reborn as the 'Ashihama Preservation Society'. The 'Ashihama' they seek to preserve is a 'Japanese' one, antedating the modern arrival of Koreans and taking as its reference the extent of the old village settlement. This preservation society acts as custodian of the history and memory of the old village. One of their recent activities involved the refurbishment of a local commemorative site for 'the oldest bridge in Japan', a claim based on the passage in the Chronicles of Japan (see Hester 1998). With funds from the city government, the site has been redesigned as a commemorative park and a designated site on Osaka's official Historical Site Promenade. The preservation society also takes a central part in organizing the twice-yearly neighbourhood festivals. Membership in the preservation society requires the recommendation of two current officers. There were no Korean members at the time of research.

The Korean-identified 'Ashihama' also remains in current use, appearing not only in daily conversation but in memoirs, novels and sociological treatises to identify Osaka's largest spatially concentrated Korean community. A more geographically localized version of this latter Ashihama also exists, denoting the visible, ethnically marked area of Korean concentration: the area cross-cutting the main local shopping street. Linking the Japanese and Korean versions of 'Ashihama', both geo-symbolically and, to a degree, socially, is 'Amatsu-dōri' shopping street.

'Amatsu-dōri' shopping street

'Amatsu-dōri' ('Heaven's Harbour Street', also a pseudonymous place name), the long-standing name of the shopping street, takes its name from the village shrine of Ashihama. The westernmost extent of the street forms the shrine's northern boundary. From the shrine, the street runs three blocks – some 500 metres – east to a canal, dug by mostly Korean labour in the earlier 1920s as part of a project to straighten and contain a local river that was prone to flooding. With the population of the area rapidly increasing, a public market was constructed in 1926 just beyond the northern extent of the old Ashihama village settlement. This served as the anchor for the development of the shopping street, which progressed from west to east.

Since the 1920s, Korean pedlars were also present on the street, mostly older women selling vegetables from straw mats spread out on the ground. This group was later joined by those selling from stalls built in a narrow alley behind the main street. By 1935 the Korean market in Ashihama was one of three large Korean markets in the city, with merchants selling from 40 or 50 shops and a further 150 or so selling from stalls or mats on the street (Kim 1985: 143–4). Korean shops apparently began moving from the back alley onto the main shopping street during the Pacific War, a process that accelerated during the post-war period. One author notes that in 1951 the Japanese head of the shopkeepers association attempted to push through a proposal to provide capital for Japanese merchants to establish shops on Amatsu-dōri in order to stem the movement of Koreans onto the street. The proposal apparently failed to win sufficient support among the other shopkeepers (Yoshida 1996: 123). Continuing expansion of the ethnic Korean community in the area during Japan's period of high economic growth, in addition to increased migration of Koreans from Cheju, enhanced opportunities for merchants in the ethnic Korean economy.

The visible face of difference

Amatsu-dōri has come to represent the visible face of a largely invisible Korean minority living in the Ashihama area. Koreans, as one of my informants once put it, 'have the same faces as Japanese'. Most use Japanese names in their daily lives.[3] The vast majority are second-, third- and now fourth-generation, Japan-born, native Japanese speakers. Nevertheless, unless they take active steps to undergo formal naturalization, these Koreans are legally foreigners.[4] So, along with periodic ethnic festivals, the marketplace displays the designs, colours, aromas and flavours of cultural difference which are the most prominent reminder of a 'foreign presence'.

Along its 500-metre length, Amatsu-dōri now hosts some 120 shops divided into three blocks, each with its own separate merchants association (*shōtenkai*). Just over half of the shops are owned and operated by Koreans, most of these handling ethnically marked Korean goods. These shops include those dealing in foodstuffs, such as kimchi, *chijimi* (batter fried with various ingredients), rice cakes (*ttok*), dried fish and meat for *yakiniku* (grilled meat), as well as those selling various dried pepper powders for making kimchi and other spicy Korean dishes; shops for bedding and ethnic Korean clothing, most particularly the traditional women's *chima chogori*; as well as a couple of Korean restaurants. There are also several Korean-run coffee shops on the street, for the most part ethnically unmarked, and two commercial financial institutions, aligned, respectively, with the ROK-affiliated *Mindan* and DPRK-affiliated *Chōsen Sōren*.[5] Providing most of the ingredients for daily sustenance of an ethnically marked Korean lifestyle, the street has long been known among local Koreans as *Chōsen Ichiba* – literally, 'Korean market' in Japanese. As well as being a symbol of the Korean presence in Japan, the street has served as a material 'ethnic' environment within which Korean residents can maintain a degree of cultural tradition and reaffirm consubstantial links to the Korean nation.

Alongside the Korean-run shops on this street are also stores that can be found on any Japanese shopping street – the rice dealer, the sake dealer, vendors of noodles or fried foods, drugstore, furniture store, barbershop and so on. Like the shops themselves, the clientele is a mix of Koreans and Japanese. However, the greatest number of customers are Korean, reflecting local demographics and the street's particular product mix.

Reviving the shopping street

The shopping street or district (*shōtengai*) is a structural element of any city or town in Japan. They are, in fact, recognized and regulated in Japanese law. While varying in size and composition, they possess certain common features that make them immediately recognizable as a social form and a type of place. Shopping districts are organized. Merchants assemble and, through an organizational hierarchy, pursue common interests. They are named, most often, after a local place. And they are almost always marked with a signboard of some sort at each entrance and by uniform lamp posts displaying signs bearing the name of each shop. Most shopping streets change decorations with the season, often displaying plastic cherry blossoms in the spring and red and gold maple leaves in the autumn. Other collective efforts include periodic sales in which all shops participate; issuing of coupons; sponsoring advertising inserts in the local newspaper; providing environmental music to shop by; and so on. The dual purpose of merchants associations are thus to pursue collective efforts to obtain economies of scale otherwise unavailable to small-scale retailers and to offer a distinctive and pleasant shopping environment to attract customers.

For at least the past couple of decades, however, shopping streets all over Japan have been suffering from chronic decline, as they face competition from convenience stores and supermarkets in such features as price, variety of goods and convenience of location or hours of operation. Suburbanization has also led to a smaller customer base for many urban shopping streets. This has led a number of shopping streets to pursue efforts to boost their appeal to customers.

The transformation of Amatsu-dōri or Chōsen Ichiba into 'Korea Town' was one such measure. But involving, as it did, the issue of Korean–Japanese relations in practice and in theme, the Korean 'theming' of Amatsu-dōri had implications well beyond the commercial health of the street. This was recognized from the first in the conceptualization of the project, particularly in the language of civic values within which the project was presented.

The Korea Town Concept

The Korea Town Concept originated with the resident Korean Osaka branch of the Jaycees, a volunteer service organization, and was worked into a formal plan jointly with the Osaka Japanese branch of the organization. It was first proposed to another, more recently developed area of Korean commercial concentration around a nearby train station. Those merchants, however, for a variety of reasons, were unable to move forward with the plan. The plan was then brought to Amatsu-dōri, where movement towards its implementation was taken up by some of the local merchants.

In its initial conception, the concept was quite ambitious, involving the marking of a wide swathe of territory with gates and various food and shopping zones, construction of new parking facilities and so on. This was quite beyond the capacity of the Amatsu-dōri merchants. But there was enough interest among some of them to move forward with negotiations within and between the three merchants associations. What was finally decided on and implemented in two of the three blocks was a 'Koreanization' of the main collective markings of the shopping street. This involved the laying of new, coloured paving stones, installation of new street lamps and attached shop signs, and the construction of archways at each end of the two participating blocks. Emblazoned on the archways were the new names bestowed on the respective blocks: 'Korea Town' in one case and 'Korea Road' in the other. I discuss the reasons for this lack of consistency in the following sections.

The basic appeal of the Korea Town plan and the marketing of the plan to the merchants was based on three basic elements. The first and most basic appeal, directed to the common interests of merchants on the street, was the prospect of the commercial reinvigoration of the street. Along with the decline facing many shopping streets, the Korean shops on Amatsu-dōri were also facing competition from the growing concentration of Korean businesses around the train station, where, in fact, a number of merchants from the Amatsu-dōri area had moved or opened up branches. A second appeal of the Korea Town project was the prospect of receiving public subsidies for the construction and promotional work it would entail, as special funds were being made available for the 'internationalization' of shopping streets. Finally, the Korean Town plan was based on a newly circulating civic ideology that made a public good of the kind of 'international cooperation' that the plan would both entail in its negotiation and demonstrate by its realization. This final element warrants some elaboration.

Towards a 'society of living together (with difference)'

Koreans are a central 'test case', as it were, for how the Japanese state and society deals with the foreign 'other', as they have been the earliest and by far the largest group of migrants in Japan's modern period. Policy during Japan's colonial occupation of Korea basically aimed to erase among Koreans attachment to Korean traditions and practices, to pre-empt identification with the Korean nation and turn the population into 'children of the Emperor', and to mobilize them towards the objectives of the Japanese state. For much of the past half century, Koreans in Japan have been placed in a position by which they were simultaneously pressured to assimilate, through official policy as well as local social norms, and socially excluded. When their cultural difference was recognized, it was usually only to be stigmatized.

The past fifteen years or so, however, have witnessed substantial changes in the way issues of national/cultural difference are addressed within Japanese society. The concept of 'internationalization' first emerged in the 1970s as a covering slogan concerned with changes in foreign and domestic policy, organizational structures and the educational system thought to be necessary for Japan to assume its proper role on the world stage. The mid-1980s saw the rise of the concept of 'internal internationalization' (*uchinaru kokusaika*), with an explicit focus on attitudes and values towards cultural difference and skills in interaction with foreigners (see, for example, Hatsuse 1985). This period also witnessed the increasing flow of a new migrant labour population into Japan in response to severe labour shortages in the booming Japanese domestic economy and a high-valued yen. With a significant number of such workers beginning to appear in Japanese workplaces and in Japanese neighbourhoods, a very prominent public debate ensued from the mid-1980s into the early 1990s over the merits of Japan becoming an open country (*kaikoku*) or remaining a closed country (*sakoku*) with regard to foreign unskilled labour. The 'internal internationalization' concept was brought into the debate by those urging a posture of acceptance of foreigners into Japanese society.

This debate posed the challenge of accommodating the culturally different foreigner on native soil – in local communities and in schools, as well as in workplaces – in a society many had long taken for granted as being homogeneous.[6] Within the context of this debate, writers and activists long concerned with the Korean issue pointed out that the recently arrived foreign workers did not represent Japan's first experience with foreign immigrant labour, as many engaged in the debate seemed to have assumed. By the end of the Pacific War, over two million Koreans lived in Japan, the great majority of them comprising an immigrant working class at the bottom of the Japanese social hierarchy. Those who remained after the war and their descendants were the principal targets of Japan's domestic policies towards foreign residents over the next four decades or so. In contrast to the 'newcomer' (*nyū-kamā*) foreign workers, resident Koreans came to be referred to as Japan's 'oldcomer' (*ōrudo-kamā*) foreigners (for example, Pak 1993). But what was emphasized in bringing these two groups into the same framework of debate was the necessity for Japanese to learn to live with cultural difference on the basis of equality and human rights, and to adjust attitudes, values, and social and legal institutions to facilitate this.

This concept of living with difference had by the early 1990s come to be crystallized in the term *kyōsei*[7] ('living together').[8] The term is by now in widespread use by resident Koreans and Japanese scholars, politicians and activists, regarding relations between the groups. In addition to an ideological framework for social relations conceived in terms of cultural difference, some local governments refer to *kyōsei* as a rationale guiding efforts to incorporate 'newcomer' foreigners into local communities through language training and other special service programmes (see Pak 2000).

Kyōsei has become a broadly sanctioned notion, especially at the level of local government and in progressive and minority circles. The concept itself contains no particular policy prescriptions regarding treatment of Japan's foreign population. What it does do, however, is draw attention to the fact that difference exists within Japanese society, not a small shift in itself, while valuing positively efforts to interact with and accommodate the foreigner in Japan.

The unfolding of the Korea Town Plan

The Korea Town plan, when it was first brought to the merchants of Amatsu-dōri in 1986, was conceived under the rubric of 'internal internationalization', perhaps one of the first practical projects to employ this language. As it unfolded, the plan was incorporated into the newly emerging civic ideal of 'living together', giving the project a value beyond simple commercial interests. Korea Town would contribute not only to the shopkeepers whose profile would be raised in a more distinctive and beautiful shopping area, but would also project a positive image of Osaka as an 'international city' and demonstrate the value of 'living together'.[9] While, realistically, Koreans in the greater Osaka area were the principal targets of the project, promoters also hoped to lure Japanese customers attracted by an 'exotic', but safe and clean, atmosphere.

Over several years, a series of meetings (to which all three Amatsu-dōri merchants associations were invited) was held on the concept of 'a society of living together (with difference)' (*kyōsei shukui*). University professors involved in issues concerning Koreans in Japan and officials of Kobe's Nanking-machi

Chinatown were among those invited to discuss this topic and share their advice with association officials. While there was a degree of communication among the three blocks of Amatsu-dōri concerning the plan, each block exhibited a largely independent response to the proposal. I would thus like to take up each of the three blocks in turn.

Within A Block, the section of the shopping street abutting the shrine, proponents of the plan were unable to muster enough support among their fellow merchants to secure the association's participation. This section of the street thus maintained its distance from the movement towards Korea Town and today bears no markings of Korea Town affiliation. As reasons for this lack of support, I was told that the main problem was a financial one, as A Block has fewer shops and thus fewer resources than the other two blocks. In other conversations, however, I was told of opposition to the idea of marking the block as 'Korean'. This block, where a slight majority of the merchants are Japanese, retains symbolic associations with the old Ashihama Village settlement. Bordering the shrine, during the summer and autumn shrine festivals, street stalls are set up along one side of the street. Having part of the festival take place under the banner of Korea Town was apparently not an acceptable prospect to some Japanese community leaders in the area. Nor was the idea of having part of the old village come under such a banner.[10]

The plan proceeded most smoothly on C Block, the block at the opposite end of Amatsu-dōri from the shrine. With the president of the merchants association of the block, a Korean restauranteur, acting as a leading proponent of the plan and with members apparently persuaded of the benefits to be gained by going forward with the plan, its implementation proceeded more or less as envisaged.

On B Block, however, the plan ran into some stumbling blocks. A couple of years after the proposal had been brought to the Amatsu-dōri merchants, a newspaper article on the plan appeared in one of the major daily newspapers. Based on information fed to the press – not by any of the merchants associations but by the originating volunteer service organization – the article was accompanied by an elaborate map depicting a Korea Town plan that incorporated a major portion of the ward in which the street was located. Some local Japanese readers reacted with outrage, expressed through a flurry of threatening phone calls to the then-president of the B Block merchants association, a Japanese man. The gist of the calls was that this area was not Korea Town, it was Japan, and any plan to make it into Korea Town had best be abandoned.

In addition to the tension these phone calls caused within the association – for the president in particular (the stress apparently affected his health) – they forced new considerations onto the block as to how to proceed with the Korea Town plan. Discussions among the merchants turned to the name Korea Town itself and what this might have suggested to their disgruntled Japanese neighbours. The Korea Town appellation had been part of the original proposal before it arrived at Amatsu-dōri.

The officials of the shopkeepers associations were not intending to make a claim to any substantial part of Ashihama for Korea Town. For the time being, they were in the midst of a discussion of a project only for the shopping street, to be based on agreement among members of each merchants association. And since the central purpose of project was to draw positive attention to the street, not to antagonize neighbours and cause an uproar, the association felt the need to respond to the concerns of the callers, even if they had been expressed in an obnoxious manner. Discussions were launched to come up with a name that would be less offensive to those sensitive to some perceived symbolic territorial ambitions among local Koreans. The long-standing popular name for the street (more commonly used among Koreans), 'Chōsen Ichiba', was put forth for consideration. Since the word *Chōsen* in Japanese has strong suggestions of North Korea and its Korean supporters in Japan, this suggestion was promptly shot down.[11] What was finally agreed on was to change the 'Town' of Korea Town to 'Road'. This would semantically confine the Korean marking to this portion of Amatsu-dōri (street) itself, which was all that the merchants had had in mind in the first place. Another accommodation to Japanese nativist sensitivities was the decision by the B Block merchants to include the name of the shopping street taken from the shrine, Amatsu-dōri B Block, as text on the arch that would bear the 'Korea Road' markings. The arch in its final form would thus bear the markings 'Amatsu-dōri B Block, Korea Road', with the first part rendered in bold black brushstroke-style Sino-Japanese characters and the latter in gold typeface-style English alphabetic characters.

Thus, some seven years after the proposal was first brought to the merchants of Amatsu-dōri, with construction expenditure of some ¥80 million (some US$666,000), much of it subsidized by the government, the Korea Town Project was unveiled at the end of 1993. This comprised 'Korea Town' on C Block, 'Korea Road' on B Block and an A Block with no Korean markings, but sporting newly installed lamps and lamp posts.

Korea Town as proscenium arch

The project was not completed with the physical changes on the shopping street. The face-lift has also been actively and creatively marketed to attract people to the street. In keeping with the theme of 'living together' with cultural difference, the street has thus been turned into a kind of proscenium arch[12] for periodic festivals and other events. These most often involve the alternating display of 'traditional national arts' (*minzoku dentō geinō*) of Korea and Japan, such as Korean *p'ungmul* percussion/dance performances and Japanese *wadaiko* drumming, which work to summarize the 'cultural' differences between Koreans and Japanese, while at the same time dramatizing mutual respect for such difference in the spirit of 'living together'. The street thus provides a social interactional space for thematizing and performing cultural difference, conceptualized as the possessions of distinctive national communities.

The politics of depoliticization

Structured by commercial interest and the need to present an attractive ambience for shopping, the promotion of difference on which the Korea Town Project is based involves something of a re-imagining in the manner in which 'Korean-ness' is symbolically deployed. As the visible face of 'Korean-ness' in Ashihama, the shopping street has symbolized resistance to the pressures of assimilation in Japanese society and all it conjures up in the historical memory of the colonial experience for Koreans. The question of what symbolic associations of 'Korean-ness' were promoted on the street emerged as another issue in the negotiation of the project.

In mid-1997 a meeting was held among younger merchants of all three associations along Amatsu-dōri. This was to be a fresh start for a coordinated effort to build on the Korea Town Project. The meeting was to centre on the adoption of a mission statement for the new Korea Town Promotion Committee that was going to be formed by these merchants. I would like to focus on one passage in the draft statement:

> In Japan, there are now three Chinatowns: Yokohama's Chūkagai, Kobe's Nanking-machi and Nagasaki's Chinatown. They have each witnessed fabulous development to become distinguished tourist spots. As regards resident Koreans, however, even though the population is proportionately greater, there is not a comparable [Korea] town. While Chinese people have from long ago had a rich experience in living all over the world, the Korean experience of living overseas has been short. Especially in the case of Koreans in Japan, there is the historical context of forced labour mobilization of the colonial period. It is a fact, therefore, that Koreans' will to settle in Japan was weak. However, having lived in Japan for half a century, resident Koreans have now begun diligently searching for the path, not of assimilation, but of 'living together' with Japanese people.

In comparing the Korean and the Chinese experience so as to bolster the case for more actively promoting Korea Town, the draft statement ventures into the territory of colonial history. This section of the statement aroused some controversy. One of the young Korean merchants stated that bringing up this issue would serve no useful purpose, since they were not engaged in a political movement but were appealing on behalf of the shopping street as a group of merchants. Factual reference to colonial history was not helpful for encouraging cooperation and could, in fact, bring trouble for the group. It was agreed that this reference would be deleted from the statement.

While this turn of events may not strike the reader as particularly surprising, it does highlight the representational logic inherent in a commercial venture and its contrast to the representational logic involved in making claims on those in positions of dominance regarding community membership or fair access to socially valued resources. In the former, the 'customer is king' and all efforts are expended on achieving consumer satisfaction. On the shopping street and among the merchants, the 'living together' framed by commercial interests is premised on avoiding unpleasantness, and thereby constrains the effort to set the project within the history of Koreans in Japan.

Another part of the draft statement cited above, however, which refers to colonial history, if somewhat obliquely, was adopted. This final passage interprets *kyōsei* in explicit contrast to assimilation. It thereby makes a claim for the retention of cultural identity among Koreans. Not so long ago, such an assertion could only have been interpreted as resistance to Japanese community and institutional norms. The language of 'living together', however, has made acknowledgement and respect for cultural difference a positive social value.

As in the mission statement itself, what is included and what is excluded, what is highlighted and what is left unspoken – that is, what is the substance to be negotiated in the ongoing accommodation involved in 'living together' with difference – will be key to how 'difference' comes to be accommodated in Japan.

Conclusion: 'living together' in the framework of national cultural difference

The Korea Town Project involved the redecoration of the shopping street with Korean-themed designs, as well as the mobilization of the Korean theme in the sponsorship of festivals and other activities to draw people – potential customers – to the street. In addition to the commercial aspect, the project was given a boost of legitimacy through what I would call a new civic ideology, summarized in the slogan of 'living together'. In this sense, the project was and is 'about' Korean–Japanese relations in both its process of implementation and in the marketing of the street to the public.

Thus the Korea Town Project may be usefully read for what it suggests about the commodification and marketing of 'culture' and the articulation of two spheres of discourse and practice – the commercial and the civic/moral – that the project called forth in its explicit theme of 'living together'. The changes in the built environment that the project entailed and the processes through which they unfolded reveal recent changes, as well as some of the tensions and contradictions, involved in the management and negotiation of cultural difference taking place in Japanese society today.

Both the Korean Town Project and the new civic ideology of 'living together', at least as manifested in Ashihama as I have observed, are efforts towards the management and rationalization of culturally ascribed 'difference'. The very name 'Korea Town', in English, and its inclusion together with Japan's 'Chinatowns' in a set of 'ethnic minority tourist sites', clearly suggests the global processes of reifying, commodifying and marketing cultural difference as one more type of product differentiation, processes of which the merchants on Amatsu-dōri are well aware and which they seek to mobilize to their advantage.

Both the 'living together' concept, especially as applied to resident Koreans in Japan and Korea Town itself in the conceptualization, rationalization and management of cultural difference, reflect a notion of culture as 'national cultural difference'. They reflect processes of categorization based on the nation-state as a fundamental cultural 'container' (Tomlinson 1991) that can be represented by such markings of 'difference' as national costume, cuisine or decorative styles. Such national categories, rendered in this way, are inadequate to describe the differences that exist on the ground in Japan between those of Korean heritage and those settled in a Japanese identity. Koreans in Japan today are overwhelmingly native Japanese-speakers, products of Japanese schools, fluent in Japanese interactional styles. Their fate is bound up with that of Japanese society and their local communities, even if they cross national borders in tracing their heritage.

The language of 'living together' seems to negotiate a third way between erasure of difference and exclusion based on difference. To the extent that it works to rationalize and manage difference based on national cultural categories, accentuating difference and assigning it to affiliation with separate national communities, this 'living together' discourse may be limited in the degree to which it can promote Korean membership in Japanese communities. This situation is reflected in the Korea Town Project itself. Local resistance to the project, based on a nativist response to the threat of a 'foreign' incursion, has left its visible marks in the landscape, such as the lack of stylistic uniformity or even consistent naming for the three blocks. Still, the fact that it was implemented at all, with the support of Japanese merchants as well as local government, is a remarkable development. This is an unmistakable sign that the evaluation of 'Korean-ness' in Japanese society has undergone a substantial change. On the other hand, this evaluation, as manifested on the street and elsewhere, is still based in national categories that tend to render local Koreans as foreigners exotic to the local landscape. In its effort to accentuate, package and market a difference recognizable within the common-sense framework of national cultural categories, the Korea Town Project seems to be caught in a contradiction between its commercial aims and the aims of local inclusion that the slogan 'living together' would seem to promise.

Notes

1 For further details on the uses of the past in place-making in Ashihama, see Hester 1998 and 1999.
2 The socially cohesive Tokyo neighbourhood reported on by Bestor (1985, 1989) was also characterized by an active neighbourhood-based economic sector in which many people worked where they lived (Bestor 1985: 128). As with Ashihama, this facilitates cross-cutting ties based in school, work, business patronage and community institutions. Compare this with the relative lack of community cohesiveness or place-making activities in the commuter neighbourhoods in Osaka Prefecture described by Applbaum (1996).
3 According to Fukuoka and Kim, in their sample of eighteen- to thirty-year-old Koreans in Japan, only around 20 per cent used their Korean ethnic names with greater frequency than they used Japanese-style names. The figure for their parents' generation was only slightly higher (Fukuoka and Kim 1997: 78–9).

4 Koreans are, in fact, naturalizing in ever increasing numbers, at around 10,000 annually. Including, along with these, the offspring of so-called 'international marriages' between Japanese and resident Koreans (who generally take Japanese nationality), the number of 'Japanese of Korean heritage' continues to grow. They remain, however, largely unrecognized. While these latter reflect the increasing integration of Koreans into Japanese society, they also present challenges to the category of 'Japanese' and its assumption of an isomorphism of language, cultural competence, lineage and ethnonational identity.
5 The *Chōsen Sōren*-affiliated banks (actually credit unions) in Osaka Prefecture have gone bankrupt. However, the Cheju-style building, of white stone design, of the Amatsu-dōri branch remains.
6 For a discussion of recent efforts of local governments to make such accommodations, see K.T. Pak (2000).
7 Kim reports that while newspaper articles mentioning *kyōsei* numbered 35 in 1989, the figure rose to nearly 200 in 1992 and almost 300 in 1997. He also points out that the term has been employed in reference to 'living together' between the genders, between the disabled and non-disabled, and in reference to the issue of Japan's former outcast Burakumin (Kim 1999: 32–43).
8 Literally, *kyōsei* means 'symbiosis', and is in fact used in this sense in the ecological sciences. In its current, more popular sense, it echoes the Japanese phrase *tomo ni ikiru* (literally, 'living together'), which is written with the same Sino-Japanese characters as *kyōsei*. I gloss the term as 'living together (with difference)', as 'difference' is the central condition to which it alludes.
9 Also part of the social context making the Korea Town Project thinkable, which it would not have been just a few years earlier, was a surge of Japanese interest in things Korean occurring from about the mid-1980s. This was brought about by widespread coverage of the dramatic movement for democracy in the ROK, Seoul's selection as host for the 1988 Summer Olympics and a general perception of the ROK's rise as one of the premier 'newly industrializing countries'. This led to what has been called a 'Korea Boom' in Japan. Accompanied by a new fashion for 'ethnic foods', especially spicy Asian food, this contributed to a 'yakiniku boom' – that is, a widening popularity of 'Korean-style' grilled meat among young people in Japan. All these might be taken as aspects of Japan's 'internationalization' (see Asakura 1994: 73–80; Nomura 1996).
10 While the shrine includes Korean parishioners, including a sizeable number who donate money, it is under tight local Japanese control, as are the local shrine festivals, even though a great number of local Koreans participate. The shrine, in fact, is a central base and a principal ideological bulwark of nativist Japanese sentiment and practice.
11 North Korea is rendered in Japanese as *Kita Chōsen*, while the Republic of Korea is known as *Kankoku*. In the mass media, resident Koreans in Japan are usually referred to as *Zainichi Kankoku-Chōsenjin* or 'Japan-resident *Kankoku* and *Chōsen* people'. The lack of an inclusive, neutral term for the Korean people in Japanese is among the reasons for the choice of the English word 'Korea' in Korea Town. The Japanese term *Korian*, derived from the English and rendered in the katakana syllabary, is now in common usage among those concerned with the issue.
12 Separating the stage from the audience, the proscenium arch is an element of theatrical architecture framing the action within as the dramatic performance.

References

Applbaum, Kalman (1996) 'Endurance of neighborhood associations in a Japanese commuter city', *Urban Anthropology and Studies of Cultural Systems and World Economic Development* 25(1): 1–39.

Asakura Toshio (1994) *Nihon no yakiniku, Kankoku no sashimi* (The Yakiniku of Japan, the Sashimi of Korea), Tokyo: Nōsangyoson Bunka Kyōkai.

Bestor, Theodore C. (1985) 'Tradition and Japanese social organization: institutional development in a Tokyo neighborhood', *Ethnology: An International Journal of Cultural and Social Anthropology* 24(2): 121–35.

—— (1989) *Neighborhood Tokyo*, Stanford: Stanford University Press.

Fukuoka Yasunori and Kim Myong-su (1997) *Zainichi Kankokujin seinen no seikatsu to ishiki* (The Life and Consciousness of Young South Koreans in Japan), Tokyo: Tokyo Daigaku Shuppankai.

Hatsuse Ryūhei (1985) *Uchinaru kokusaika* (Internal Internationalization), Tokyo: Sanrei Shobo.

Hester, Jeffry T. (1998) 'Walls and bridges: narrating the nation in an Osaka neighborhood', paper presented at the American Anthropological Association meetings in Philadelphia, 2–6 December.

—— (1999) 'Place-making and the cultural politics of belonging in a mixed Korean–Japanese locale of Osaka, Japan', unpublished Ph.D. thesis, Department of Anthropology, University of California, Berkeley.

Kim Ch'an-jong (1985) *Ihō jin wa Kimigayo-maru ni notte* (Foreigners Board the Kimigayo-maru), Tokyo: Iwanami Shoten.

Kim Tae-yon (1999) *Aidentiti poritikusu wo koete* (Beyond Identity Politics), Tokyo: Sekai Shisōsha.

Nomura Susumu (1996) *Korian sekai no tabi* (Korean World Travel), Tokyo: Kōdansha.

Pak Il (1993) 'Ethnic minority in Japan: Korean residents' case', *Keizaigaku Zasshi* (Osaka Shiritsu Daigaku Keizaigaku-bu) 94: 25–35.

Pak, Katherine Tegtmeyer (2000) 'Foreigners are local citizens too: local governments respond to international migration in Japan', in Mike Douglass and Glenda Roberts (eds), *Japan and Global Migration: Foreign workers and the advent of a multicultural society*, New York: Routledge.

Robertson, Jennifer (1991) *Native and Newcomer: Making and remaking a Japanese city*, Berkeley: University of California Press.

Sugihara Toru (1998) *Ekkyō suru tami: Kindai Osaka no Chōsenjin shi kenkyū* (People Crossing Borders: Research on the history of Koreans in modern Osaka), Tokyo: Shinkansha.

Tomlinson, John (1991) *Cultural Imperialism: A critical introduction*, London: Pinter.

Yoshida Tomohiko (1996) 'Nihon no toshi ni okeru gaikokujin mainoriti no teijū kankyō kakuritu katei ni kansuru kenkyū' (Study of the process of settlement establishment by foreign residents in Japanese cities), Toyosaki-shi: Self-published.

12 The appropriation of public space as a space for living
The Waterworld Festival in Vienna

Heidi Dumreicher

The way in which public and private space is used, organized, shaped and furnished expresses societal coherence, as well as the meaning this space has for individuals. I understand the 'shaping of public space' as the outcome of the interaction between individuals and societal groups on a spatial dimension. The analysis of a multi-ethnic street festival in its local and regional dimensions will show whether such an endeavour can open up *Möglichkeitsräume* ('spaces of possibilities') for multi-ethnic co-existence in a city, not only during the festival but also in its day-to-day public life.

Public space is made up of a physical and a social environment; the built environment is, then, a mould in which social activities take place. Both factors – the physical and the social aspects of public space – influence each other reciprocally and mutually in an iterative process. In our era of globalization, public space is the spot where a societal process of negotiation between local dwellers and immigrants, between the well known and the unknown, between the powerful and the powerless can find its public stage.

The newcomers look *fremd* ('foreign' or 'strange') to the locals. Parallel to the withdrawal into privacy, specific to our time, and being drawn into latent instead of manifest conflicts, the status of 'the stranger' has changed from being an object that is stared at to being a creature that is scientifically interesting. At the same time what is 'foreign' about the foreigner is not so clear. With globalization, it is an omnipresent similarity and equality that is favoured, not 'foreign-ness'. In the final analysis, nobody should be 'foreign' in any place; places as well as people are identical everywhere (Schütze 2000). Globalization tends to abolish the 'foreign', leaving a vacuum behind. Global forces standardize urban space and urban behaviour; if this process continues, nobody and nothing will be 'foreign' in the end. Many spots remain, however, where local features, including those introduced by foreigners, hold their own against processes such as McDonaldization (Ritzer 2000).

Vacant public spaces in Vienna and elsewhere can encourage ethnic diversity and the co-existence of representatives of this diversity. Those spaces exist, but they need people, initiatives and encouragement to become spaces for 'social know-how', for encounters and to create options now and in the future. Vehicle-free spaces are the physical conditions for a public unfolding of ethnic diversity, but the existence of those spaces alone will not suffice. Social activities must occur in order to build, furnish and use the vacant space.

The Viennese non-profit organization 'Oikodrom Institute for Urban Sustainability' (or Oikodrom for short),[1] of which the author is the founder, carries out research programmes on sustainability in densely populated regions and encourages processes of sustainability in towns and cities. These studies are all interdisciplinary, including architects, experts on reducing energy consumption, sociologists and ethnologists (Dumreicher and Levine 1996, 1999; Dumreicher *et al.* 2000; Levine, Yanarella and Radmard 1998). In one of its projects, Oikodrom is monitoring a sustainability process in a Viennese district called Rudolfsheim-Fünfhaus (the 15th District of Vienna with 70,000 dwellers). The Oikodrom team has established good contacts with the Regionalforum, a platform that represents fifty local social initiatives in this district. These initiatives are all concerned, in one way or another, with multi-ethnic interests, giving consultancy to immigrant women or to schoolchildren, besides creating multi-ethnic discussions and playgroups in the green parks of the district.

Oikodrom has contributed theoretically and practically to Agenda 21, the agenda for a sustainable world in the twenty-first century that was set at the UNCED (Earth Summit) at Rio de Janeiro in 1992, and to the Aalborg Charter for European Cities and Towns Towards Sustainability (Dumreicher 1995). Oikodrom, with such a background, has initiated a transdisciplinary process between, on the one hand, scientists with their 'disciplinary know-how' and, on the other, local experts with their 'social know-how' and tacit knowledge. At such meetings in Europe, the NGOs often tend to limit themselves to a one-way stream of words. However Oikodrom, by becoming a part of the Regionalforum, has participated in the local process of discussion and action. Such a discussion led to the *Wasserweltfest* (Waterworld Festival), a multi-ethnic cultural festival that took place on 16–17 June 2000 and attracted 12,000 people (Stadt als Bühne 2000).

In this chapter I will discuss the assumptions behind this festival, the bureaucratic response to it and the way it unfolded. It is an example of how a vacant public space can become an inter-ethnic meeting ground, albeit only for two days. As a follow-up to the Waterworld Festival, Oikodrom and the Regionalforum have created a series of monthly events, taking place on the afternoon of the fifteenth day of each month, aiming at the multi-ethnic dwellers of Rudolfsheim-Fünfhaus. Another development was the Waterworld Festival II, which took place in June 2001 and was an even greater success in the district.

The Wasserweltplatz: its individual, architectural and societal meanings

The Kardinal Rauscherplatz, or Wasserweltplatz (Waterworld Square) as it is popularly known, is a huge square, 500 metres long and 25–150 metres wide, located in the 15th District of Vienna. On a normal weekday, the square has a busy section – between an underground station and a covered market – known as Meiselmarkt. There the square is dominated by people carrying their empty or full shopping bags. To them, it is a space without any meaning other than 'getting from A to B'. People leave the underground station, head towards the entrance to the market and, half an hour or so later, cross the square in the opposite direction, again without even conceiving the idea of it as an available and usable space. For the rest of the time, this space is not used at all. Nevertheless, the Wasserweltplatz does signify something for both society as a whole and individuals.

The Wasserweltplatz expresses societal demands. Seven years ago, the city administration decided to transform this urban area with its heavy traffic into a pedestrian zone. It started to refurbish the existing old-fashioned and somewhat shabby marketplace with its numerous little booths into a modern facility – a shopping mall to suit a multi-ethnic, suburban public. The intention was to create a new synergetic point or meeting place in the neighbourhood and thereby improve societal coherence.

In addition to the public meaning ascribed to this square, the Wasserweltplatz expresses private interpretations, in particular that of the architect, Rudolf Gutmann, who was given the task of transforming the car-laden streets into a pedestrian area. Gutmann is a much-respected Viennese architect and the leader of a group of renowned international architects. He saw in this project a challenge for his own work: finding a style for a place where different ethnic groups could spend their leisure time. The district has the most intense concentration of non-Austrians and recent immigrants: 30 per cent of the population consists of *Neu-Österreicher* (New Austrians) or *Neu-Wiener* (New Viennese), as they are popularly known. They have come from Turkey, the former Yugoslavia, the Czech Republic, Slovakia, Greece, China and other countries. Through the construction of a fountain and the attraction of flowing water, Gutmann wanted to create a space into which the residents could escape from the extreme density of the rest of the district. He hoped that the site would stimulate new social processes.

But his simplistic approach to the complicated participatory process of shaping the square for use by different people did not succeed. Although situated in a district with an urgent need for space (and despite the fact that it was freed from traffic), the square is usually empty. The city has provided some benches, but their use reflects a sorry state of affairs: people only sit down if they need a rest because their shopping bags are too heavy to carry in one go. People who sit there give the impression that they have no other place to be, as if they are outcasts: Wasserweltplatz is not a place to meet your friends, even less a place to have new experiences; no one wants to seek out the shade of the trees; children do not play in any of the different water arrangements – if they did, they would probably be chased away.

Into this barren environment came the Waterworld Festival. It was created on the theoretical basis both of the sustainability studies that Oikodrom had carried out in previous years and of the 'social know-how' existing in this lively district, full of contradictions and with a lot of social problems. The partners created a platform for a Festival that was meant to increase the quality of living in the district by providing a meeting place for the residents. The organizers invited people who had shown interest in interacting with other people and in giving away something of themselves. The organizers also invited artists – from the whole of Vienna and from beyond the city – to contribute to the emerging hybrid but authentically developed culture of a suburb in a post-modern and post-industrial metropolis.

Two examples give an idea of the multi-ethnic displays that took place in the festival. The first example, from 16 June 2000, is *Das Wohnzimmer des Herrn Stern* (Mr Stern's Living Room) – 'Mr Stern' being the equivalent, in symbolic meaning, of 'Mr Smith' in English. A street-theatre actor on roller-skates arrives in the square carrying a transistor radio, heavy luggage and so on – the objects indicating that he has taken a long journey. He finds a convenient place near the underground station and starts his performance. He looks around as if thinking, 'What is this space around me?'; 'Where am I?' He opens his suitcase and takes out strange and amusing objects. In no time, the young-sters from the square are following him through the crowd; they form a single group of actors, with no distinction between the 'actor' and 'spectator', moving over the whole square. The non-verbal character of the performance attracted spectators regardless of ethnic origin. Intercultural 'discourse' is thereby opened up.

The second example also took place on 16 June: A Turkish storyteller, a young woman, summons up her courage and starts to tell stories in Turkish and German. She is a professional, but this is the first time that she has spoken in a public space without a predefined or specific audience. (She usually performs at kindergartens or children's birthday parties.) She has decided to furnish the space for her performance with a huge red piece of cotton cloth, which provides the necessary background. The event has significance for her as an individual: she has never had the experience of working with an unknown public. Her individual relationship to public space is new, and she learns that public space is not hostile but, on the contrary, inviting. She had decided to use both German and Turkish when telling her stories. Her individual choice was, in fact, a societal decision: she would attract German-speaking kids when speaking in German and children whose mother tongue was Turkish when she was speaking in Turkish. In her own person too, a plural ethnicity is constantly displayed: her outfit and accent reveal her double identity.

In short, in so far as public space consists of physical and social aspects influencing each other – as was stated previously – this interactive process and mutual relationship developed from a very rudimentary form into a more sophisticated one through the activities of the Waterworld Festival. The local activists of Regionalforum, together with the professional consultants of Oikodrom, decided to try to breath life into Wasserweltplatz by organizing the Waterworld Festival since, in the course of daily life on the Wasserweltplatz, there is very little contact between the different ethnic groups. In reality, a storyteller alone cannot overcome the invisible barriers that exist. But she can become a temporary symbol of an era of shifting identities: she is, all at the same time, Turkish and Viennese, rural and urban, traditional and twenty-first century, myth-addicted and practical. As one of the displays put on for a very big inter-ethnic event, she contributed a meaning to the Festival, while the overall concept of the Festival took on new significance for her.

Public space as a space for living: seven theses for an emancipatory urban development

With its 'Compass for Sustainability' (Dumreicher and Levine 1999), Oikodrom has laid down the basis for an emancipatory process in towns and cities, with the accent mainly on the concept of *Möglichkeitsräume* ('spaces of possibilities'). Creating spaces of possibilities is one way in which forms of human interaction can be encouraged. These spaces can be envisaged as physical ones – like the built environment – or as spaces for encounters, ideas and options in the present and the future. A festival in an open space represents both parts of this new notion: the existing, built square becomes a place for physical action, such as presentations or a bouncy castle, and it also becomes the physical basis for the encounter of ethnic groups, of visible and non-visible differences. It has often been said that it is enough to put such a space at the disposal of those who live there: by themselves, they will develop public activities. But the experience in the 15th District shows that intervention, rather than the provision of a space, is needed. Space needs people to build on it, to liberate it, to furbish it, to use it and to own it. Such was the point of departure for Oikodrom and the Regionalforum: their common hypothesis and working programme had been that space, in its architectural and urban form, was meant to encourage ethnic diversity and the co-existence of representatives of various ethnic groups.

The interdisciplinary Oikodrom team has formulated seven theses relating to the quality of a town or city (Dumreicher and Kolb 2001). They were developed using an empirical methodology – with objective hermeneutics – as a tool for sorting reality in such a way as to provide researchers, local experts and people supporting urban sustainability initiatives with a common basis for discussion and action. These theses are based on interviews with ordinary city-dwellers who gave us their views on their own city. We were able to discover common basic structures in the ways that city-dwellers of different genders and from various social and educational backgrounds perceived their urban environment. These seven theses functioned as a manifesto or a basis on which Oikodrom developed the Waterworld Festival. They are:

Thesis 1: Towns and cities are diverse. The city, in its complexity, generates dichotomies with manifold images, such as human scale versus anonymity, freedom versus loss of control.

Thesis 2: Urban life needs a town or city that provides services. The attraction of a town or city derives, in part, from the provision of basic urban services, from *Nahversorgung* (neighbourhood care) to public utilities and services. Conceptualizing the sustainable city means reconceptualizing the way in which the city administration and ordinary citizens share roles and responsibilities, without resorting to excessive moralizing.

Thesis 3: Emotional co-ownership of the city: use creates significance. Only when the inhabitants use the potential of the city (including the public spaces provided), are they able to fully identify with the place, thereby releasing it from abstraction.

Thesis 4: The city is a stage: the town is a place for self-expression. Today, this quality has been reduced by the 'tyranny of intimacy' and by over-commercialization.

Thesis 5: The city or town combines the near, the foreign and the faraway. The quality of urban life is a function of the quality of the living and working environments, in combination with the opportunities offered by the city centre, together with its integrated suburbs and its immediate surroundings.

Thesis 6: The city map in the mind represents the structure of the relation of the urban residents to urban spaces. Points for orientation in the city are, on the one hand, tourist attractions and main foreground buildings and, on the other hand, the not so spectacular places that attract and support urban life through their functional quality.

Thesis 7: Space by itself is formless, but it is shaped by objects with form. Space itself is limitless, but it gets a tangible form through the scale and the quality of physical elements. Space is determined by the ways in which it is shaped and limited by physical objects, from the scale of buildings to the scale of urban furnishing within the space.

How the seven theses relate to the Waterworld Festival in terms of public space as a space for living

As the Waterworld Festival is an outcome of Oikodrom's theoretical approach to the sustainable city, the team has used the seven theses in order to check whether the festival actually corresponds to popular needs and contributes to a participatory and multi-ethnic concept of urban life.

Thesis 1: Urban diversity

A scene from the festival: Strange creatures appear, wrapped in giant stockings, in front of the brick Catholic Church. They move slowly over the ground in front of the church – as if from an alien world, expressing themselves in human and non-human movements – forming statues, dissolving or becoming creatures that seem to come right out of the sewage system of the city. The

Viennese theatre group *Vis Plastica* is in action. This contemporary theatre performs on the space – similar to a typical village square – at the entrance to a church that usually attracts elderly Viennese people of rather conservative tastes. But in the overall atmosphere of the festival, even the avant-garde has its space and place. As one guest wrote in answer to the organizer's questionnaire: 'Das Gesamte zählt, nicht das Einzelne' (the parts do not count, but the whole does).

Since the public was actually involved in many of the stalls or displays that made up the festival, the organizers succeeded in realizing their basic concept: that the festival should be interactive. The festival had to have such dimensions as to enable diversity and fulfil very different needs and wishes. One social worker, Ilse Marschalek, remarked:

> So many things occurred in parallel that I hardly knew what was going on. The festival could have generated negative dynamics, given the difficult background in the district.

Marschalek is a staff member of Oikodrom and was appointed coordinator of the festival following her nomination by the Regionalforum.

So a loss of control, typical of many metropolitan situations, occurred; but, surprisingly enough, this was felt as being something positive, not negative. In the overall positive situation of a 'space of possibilities', Ilse Marschalek could hope and trust that people would organize themselves and use this freedom for the benefit of the whole and for themselves as well. In an era of globalization, when the 'social know-how' used by city-dwellers to find a way to deal with the urban diversity reaches its limits, events like the Waterworld Festival can help in developing citizens' ability to cope.

Thesis 2: Urban services

A scene from the festival (or rather just before the festival): *Kollaudierung* (issuing a permit) is a word that frightens everybody who wants to organize an event in public space in Vienna. The *Kollaudierung* is a ritual by which a group of town or city professionals examine your concept and raise their concerns on the spot: Is the Waterworld Festival acceptable from the point of view of safety (electricity and so forth)? Is it in accordance with the rules concerning public space? Will it cause too much noise? Will there be rubbish that the city will have to remove? Will it provoke disorderly behaviour so that the police will have to intervene? Will life in the covered marketplace suffer from the diverse activities? It turned out that the representatives of the administration were friendly and willing to negotiate, both when issuing the permit and also during the festival itself.

On the last evening of the festival, one of our guests called the police because his mobile phone had been stolen. Normally, this would have caused a serious change in atmosphere, even at a peaceful festival. But on that particular night, the policemen reacted extremely tactfully, helping to defuse the situation.

In a general context, this thesis means that the concept of 'urban services' needs redefinition. Besides traditional services like public transport or waste disposal, a new kind of service has to emerge in this age of participation to encourage the civic activities of *der mündige Bürger* (the articulate citizen). A new relationship between the administration and politicians, on the one hand, and their civil partners, on the other, appeared during the festival. We, the organizers, showed the city that we were able to organize a relatively big event by ourselves, thereby forcing the administration to overcome their initially rather hesitant approach to our proposal. From the usual obstructive tactics and attempts to make things impossible – a game that bureaucrats play very well – they moved to a position of goodwill. The strategy employed by non-governmental organizations had its idiosyncratic Austrian style: if one of their representatives managed to break down the bloc mentality of the bureaucracy and negotiate at an individual level, they might succeed in getting their point accepted. In our case, it was possible to proceed by engaging the help of people from within the administration. Many unconventional ways of doing this were brought to light: Sometimes we were told to apply to some public entity under different names, sometimes we were directly supported by civil servants who worked for 15th District in dealing with other city officials, and so on. In a situation where the bureaucracy is normally very hard to deal with, these individual solutions were the only way to succeed. Although this is a familiar phenomenon for all activities that start at the bottom, it is, of course, not a way to approach things for the future.

'Service', in our terms, does not mean just basic services. This has been more or less fulfilled at Wasserweltplatz: the neighbourhood services work well and they have even provided the residents with a well refurbished public space – the square itself. But what is missing is the connecting 'software': the town or city must also provide non-material services.

The architect Rudolf Gutmann had constructed a good place for the establishment of a restaurant that would, at the same time, have facilities – like, for instance, an electrical outlet that could be used for outdoor music – for public events on the square. But this restaurant has turned into a 'ghost' project, because the proprietor failed to reach an agreement with the city about the way he should operate, which led to the famous standard answer: 'Nothing can be done about it.' So instead of becoming the planned communications centre, the restaurant on the square is now a black hole – a 'place of dishonour' as the owner of a little shop behind the establishment put it.

The solution is not that the public sector should take over the restaurant, but rather that the discussions with the proprietor should be brought to a conclusion in such a way as to satisfy the expressed needs of the people. The 'urban service', in this case, consists of taking the initiative in negotiations with the proprietor of the restaurant, encouraging and reducing bureaucratic procedures. Such public–private interest in representing the needs of special groups of the population seems to be more and more important, especially in a district with a large degree of ethnic diversity.

At one event, Oikodrom too experienced the opposite of this envisaged tolerance. A film show at Wasserweltplatz ended at 22:30 instead of 22:15 as arranged, and Oikodrom's executive manager faced a criminal charge. She was ordered to pay a fine of €600 or serve a seventy-two-hour prison sentence ...

Maybe the new interpretation of 'services' seems a luxury that only the developed world can ask for, since basic needs like sewage disposal have already been taken care of. But we feel that social needs are just as important as basic needs and they might well be better appreciated in the so-called 'developing and threshold countries'.

Thesis 3: Emotional co-ownership of the city

A scene from the festival: A *Luftburg* (bouncy castle) was the *big* event that many of the kids of various ethnic backgrounds had been looking forward to – in no time 'jumping' turned into a common language that crossed all cultural barriers. At the end of the day, all of the kids had had a good time. Even when the castle was deflated and jumping was no longer possible, they still stayed and combined their efforts in pressing the remaining air out of the castle and packing it up. The Yugoslavian owner of the castle, Mickey, stated that the children were able to play, talk and laugh in this part of the public space because it was an emotionally, not economically, shared space.

'Die Stadt gehört dir' ('The City is Yours') is a slogan launched by Vienna city officials. This slogan is seen on posters in many places, such as on the underground. But the *dir* (yours) in fact addresses people in the singular,[2] implying that the individual is the owner of this city, that the city can be considered a sort of supermarket from which one can take what one needs without having to pay.

The Waterworld Festival, in opposition to this notion of individuality, was a process of 'co-owning' – of living and experiencing a new city history. It was a festival of the self-constituted Regionalforum, a free association of all the groups that have undertaken social work or developed initiatives in the district in question. The Regionalforum had met monthly for many years, but had never developed a common ground for large-scale action. Instead, they had done their work quietly and anonymously. When they were preparing for the event together, however, they began to see themselves as owners of the city and not as marginalized groups pursuing certain social initiatives. In working together (that is, co-owning the public space), they experienced the synergistic effects of joint work.

This found its expression in spatial form. It became clear that for two days the Wasserweltplatz was the people's own territory. It was also clear that this explicitly co-owned territory needed large spatial symbols to indicate this. The organizers of the festival developed completely new furniture on a grand scale: the bouncy castle; a marquee put up by the scouts where guests could experience a journey into the unknown; a huge painting placed at an underground station that could be seen from at least a kilometre away; paper decorations that formed bridges over the square by binding the streetlamps, roofs of buildings and trees together, besides the props of *das offene Wohnzimmer* (the Open Living Room). Both participants and organizers had a strong sense of this territory; for instance, when leaving the square in order to make a telephone call, it was clear that one had crossed an invisible border.

The large size of the props also put the support of the administration to the test. The huge painting, which had been made by a sixteen-year-old girl over several months, had transgressed invisible borders between what the administration usually permitted and what they ordered to be removed.

The fact that the whole event was not a commercial but an emotional event led to manifold positive effects. It was *big*, with more than 12,000 visitors to a square that is usually empty, and yet there was space for everyone to make their own contributions. We found specialists for noisy tasks, specialists at accompanying VIP guests and policemen who could provide links with the city administration.

The bouncy castle became the project of a resident of the neighbourhood whose desire was to transform the yearning of the street children into reality. This conforms to Augusto Boal's concept of 'desire' as being the driving force for participation (Boal 1995). The term 'desire' is used here as a translation of the German word *Verlangen*, but *Verlangen* also means to ask persistently for something until you get it. The poor man had to work hard to achieve his desire: The electrical connections would not fit and he had to struggle with the authorities for more than two hours over his intentions – with a bunch of young kids around his legs, giving physical support to his aims.

During the festival, different places provided various possibilities for emotional co-ownership. The youth groups had an extremely noisy section. A nightclub, *Blue Tomato*, created a little bar at the edge of the square that became a meeting place for all the 'actors' – and a refuge for the tired organizers.

For two days, the inhabitants – old as well as young, immigrants as well as long-term residents ... and the Regionalforum itself – owned a section of the city. During the festival, the social form of sustainability had found its place through 'localization', an important aspect of a sustainability process. The day after the festival, however, was a bitter experience for the kids, who had experienced the city in its lively form only to see that certain activities were again forbidden. 'When will you be back?' asked the youngsters, present on the square after the festival in larger numbers than before, of the members of the team of organizers whenever they happened to meet one of them. They especially asked this of Ilse Marschalek, who had become a symbol of peace, freedom, multi-ethnic co-existence and fun.

For two days, the immigrants had experienced emotional co-ownership of 'their' urban space and had been active participants in 'their' urban life. Through a symbolic but non-destructive act, they performed a sort of ritual, defining the limits and the borders of the multi-ethnic event they had just experienced and loved. They showed that they were aware of the contingent character of this event; it would not last forever. So the event was given a precisely defined ending. Afterwards, their experience of social exclusion found expression and became visible again.

The Vienna city administration, as well as Viennese journalists, use the neologisms 'New Austrians' or 'New Viennese' to distinguish the different types of immigrants. These are euphemistic terms, avoiding the usage of 'immigrants', 'foreigners' and so on. It makes no distinction between different ethnic groups when one is trying to avoid being too specific while speaking about immigrants. The term 'Neo-Austrians' refers to them all: the legal and the illegal, the immigrants who already have an Austrian citizenship and the ones who have been living here for a long time without achieving the bureaucratic sanction of becoming naturalized Austrians. It also includes 'visible' ethnic minorities (such as conservative Muslims, Asiatic ethnic groups and so on) alongside the 'non-visible'. The term 'New Austrians' prevents endless discussion of the correct use of terms, discussion that tends to distract attention from the real issue: What kind of integration do people want?

The 15th District is one of the most multi-ethnic neighbourhoods of Vienna, where about 30 per cent of the residents are New Austrians. The aim of the Waterworld Festival was to find a way to offer New and Old Viennese two days in which to intermingle in a public space. The guests of, visitors to and participants in the Waterworld Festival were 50 per cent Old Austrians and 50 per cent New Austrians. We, the organizers, were happy that the 30 per cent minority of New Viennese had contributed as many as half of the number of guests. However, this also meant that a smaller proportion of the 70 per cent Old Austrians in the district had crossed the border to this temporary multi-ethnic event. For most of those who accepted our invitation, it was the first time that they had been in contact with their New Austrian neighbours and fellow citizens.

Thesis 4: The city as a stage

A scene from the festival: At the exit of the Meiselmarkt, the covered marketplace, people who have finished their shopping usually climb a set of stairs in order to get to the square. Right beside the exit, the festival organizers set a professional stage where artists, youngsters, musicians and fashion models could perform. The shoppers, instead of leaving as usual, suddenly find themselves to be the 'actors'. The public has spontaneously transformed the curved set of stairs into an 'amphitheatre' facing the market, so that whoever attempts to come out of the market is directly in sight of this impromptu audience.

Thus to see and to be seen – the traditional use for urban public space – has become a reality for the residents of this district, notwithstanding 'the tyranny of intimacy' that Richard Sennett (1994) has described. Sennett suggests that this tends to drive people away from public space in post-industrial societies – away from the street and back to the TV, to the internet, to the phenomenon of being 'cocooned' so that activities inside the house become more important than activities outside. During the festival, the mere passive act of seeing, from inside the house, the different possible 'scenes' emerging outdoors, turned into a sort of active seeing: One had to go out in the street to look at a performance or a musical presentation. What is more, at the same time one could participate in the event and could be seen. This two-way aspect, which is an important feature for people in public space in general, turned out to happen often at the Waterworld Festival. The borders between acting and reacting were not precise. From 'being seen', people could – if they wanted – very easily convert to being people who 'observed'.

The kids – from eight to sixteen years old – who had participated in the preparations for the festival seemed interested in the events and the stalls or displays that the organizers had set up. It was equally interesting for them to be provided with a stage where they themselves could be the actors. They presented their music and their fashion show, invading the existing stage and catwalk, with spectators watching them. When the official stage did not leave enough time for them between acts, they created their own stage: They climbed on walls and fountains, dancing and singing and offering a wonderful breakdance performance.

Places that normally just functioned as space for doing something also turned into a special stage. The children used to play basketball near a church close to the square. During the festival, when people started to take snapshots of them, it was not clear any more whether this was part of the official programme or 'just some kids playing basketball'. One can see from their photographs that these kids were perfectly aware of being taken to be comedy actors due to their behaviour. Their motto: Play basketball – and be seen when playing.

The way in which the guests of and participants in the Waterworld Festival behaved seems to be a very twenty-first-century version of a dandy prancing through the city centre. In a suburban situation, with a population of low social status, people promenaded – walked in a festive way – in their specific ethnic attire. This was the case for a couple in evidently Indian clothes, with the woman wearing a remarkably beautiful sari. It was the case for youngsters riding a bicycle in order to attract the attention of their friends. It was the case for Turkish couples with the women dressed in what one would call 'variations of the Turkish style':[3] some in the traditional *shalvar* (baggy trousers), long-sleeved shirt, knitted vest and *yemeni* (a cheesecloth headscarf, white or with a printed design), while some had taken to wearing a large colourful scarf and an ankle-length overcoat, along with opaque stockings, a consequence of the fashions created by the fundamentalist movement. This was also the case for some Turkish women who insisted on wearing skirts over trousers, a long-sleeved blouse with a cardigan over it, and a headscarf covering the head – labelled as 'concierge fashion' in Turkey. In contrast, the more urban and modern Turkish women preferred to dress up as Western urban women, though some covered their heads loosely with a scarf.

The town or city as an urban stage is present in collective memory. Self-representation is necessary, because otherwise you cannot be seen and understood. Today, under post-modern urban conditions, people who meet strangers cannot 'interpret' them, for their personal histories and origins are unknown and hard to discover in an anonymous world. Visible cues about one's identity become important in the social interactions of strangers. Hence, the provision of opportunities to display one's identity becomes an important quality of the modern town or city.

Usually commerce is the driving force in events like street festivals. Vienna excels at organizing such events, like big opera performances in front of the Rathaus and preparations for the grand opening of the *Wiener Festwochen* (the official Summer Festival). These events are part of the city administration's strategy to attract tourists, who will then spend money and strengthen the economic life of the city. At the *Wasserweltfest*, the economic aspect also played a role in the form of cooperation between social initiatives and the local market, the restaurants, the cafes and shops – a cooperation that rarely happens. Their aim was to strengthen the local economy and not the economy of the city at large, with the intention of building up a local identity that included and negotiated the different interests present with relation to this public space.

In most street festivals in Vienna, only a few actors are heard and seen. It was thus important that the Waterworld Festival opened the city-as-stage up to more players than could usually perform. This was all the more important because the city, as a stage for protest, has become part of the emerging global civil society: protests against discrimination, against the World Bank, against airports and so on are part of the normal inventory. But lately the NGOs have shifted from protests against things, to protest in favour of things, enjoying a rather unconventional moment together in the presence of the public.

Thesis 5: The near, the foreign and the faraway

A scene from the festival: The last evening. It was getting dark. Everybody was tired. We were in the 'red-light district' of a zone of the city where there is latent xenophobia. It is also a district of underlying – and often outspoken – conflicts between different generations, manifested especially in parks and public spaces. A negative gust of anger was about to invade the festival. Youngsters destroyed 'the Open Living Room'. A man had his mobile phone stolen and a group of policemen had appeared on the scene. Another man had severely cut his arm – blood gushed from his wound. The ambulance with its loud siren seemed to put an end to the friendly and open atmosphere that had prevailed over the last two days. The police had become concerned about our last and late-night programme: the Brazilian *capoeira*, including fire-breathing. But these strangers, the *capoeira*-dancers with their struggle-for-liberation ballet, which seemed so unsuitable for a Viennese suburb, turned out to be exactly what was needed. The aggressive style of the dancing took over the atmosphere in the square and changed its apparent destructive energy into a contribution to the positive energy of the event.

The metropolis is a place of many dichotomies, including, among others, the dichotomy of nearness and distance or the known and unknown. Village-like

Public space as a space for living 205

aspects give the urban resident a feeling of 'being at home', spatially and socially. The festival had these qualities: many visitors talked of feeling as if they were on a village square. The necessary solidarity-patterns evolved almost from nothing. Urban renewal movements like the Italian *Città degli villaggi* take these desires into account. The festival had this quality of a neighbourhood, but 'feeling at home' is not enough in a metropolitan situation. The city-dweller is also heading for the unknown, even if the urban neighbourhood is the place for local participation. Town- or city-dwellers are aware of being part of a bigger whole, where other interests, other cultures, other styles than their own have a *raison d'être*. In short, by making the 'foreign' and 'faraway', at least for a moment, part of the neighbourhood, the Wasserweltplatz took on greater symbolic value and became a place for multiculturalism.

Thesis 6. The city map in the mind

The mental map of the district has, perhaps, changed a little. When the administration had decided to renovate Wasserweltplatz, it intended to make it a remarkable place on the citizens' mental map. But they failed. One reason for the failure might be the fact that they were following accepted notions about the built environment – that a tower, bridge or church are the important architectural aspects of a town or city. But, at a pragmatic and symbolic level, orientation points are much more differentiated. Dynamic places – places where something is in motion and places of social change – are equally important to the mental map of the resident of a town or city. Bridges, railway stations and marketplaces are all important independent of their architectural forms.

Another reason for the failure is that citizens want the meaning of buildings to be understood. An architectural monument is important to a resident because it gives more value and meaning to that resident's neighbourhood. This effect can also be achieved by attaching a place to a specific event. Looking down on the Wasserweltplatz from the buildings around the square, some people may not have been pleased with the multi-ethnic aspect of what they saw. 'Warum bringt ihr die Tschuschenkapelle, Tschuschen san mir selber?' (Why bring the wog's music here – are we foreigners ourselves?).[4] But, on the whole, the people living around the square took the event as an adventure that showed the importance of their neighbourhood. The city and/or some artists had taken notice of it and had given the square a style and shape for a period of time. Where the city administration had failed to create a place, the Waterworld Festival had succeeded.

Thesis 7: The shaping of space by objects with form

A scene from the festival: 'The Open Living Room' (*das offene Wohnzimmer*) was created in the middle of Wasserweltplatz. It was furnished with a Turkish rug, two armchairs, a couch, a nice lamp, a table, a teapot and teacups, and a person to speak to. This display, despite being somewhat absurdly situated in the middle of a public space, was recognizable in form; a fact that was immediately understood and respected by the public. Austrian and Turkish youngsters – the ones that old ladies might be frightened of when they met them in the underground – would politely knock at the invisible door before stepping on the carpet. Other young people also participated, playing at being the mother and the auntie, serving tea and behaving as if in a real living room.

At one point, however, the organizers had to remove the carpet because another street theatre act needed the carpet as a backdrop in order to absorb the vibrations of their drum orchestra. The living room lost its perfect or usual 'shape' and thereby immediately lost its meaning. One boy took the teapot, put it on the floor and placed a big piece of paper against it which stated 'I am hungry and I have no job', starting to collect money from the participants in a part serious, part pathetic, part ironical and part self-critical way. Once the physical wholeness was gone, the 'hostess' of the living room also lost her context: From playing at being a respectable woman with strange rituals of serving tea and acceptable chatter, she changed in the perception of the participants into a mere organizer. The young people began to ask her whether she would give them the furniture at a low price – or maybe for free. She agreed, since the furniture, without its context, had become a number of disparate objects. One young man took the couch, lifted it above his head and ran away with it – a couch turning into a war trophy, maybe into a victory banner. Then a chain of changes in use began. For instance, for a while the couch – no longer part of a living room, but still useful on its own – was used as a comfortable seat from which to watch the ongoing official programme. For once the wogs (*Tschuschen*) were sitting on some sort of sumptuous balcony seat, behaving as if they were at the Vienna Opera House. Having lost its coherent context, the furniture also lost most of its meaning. At the end of the festival, we, the organizers, found the broken pieces of that beautiful couch, which had served different purposes throughout the two days, thrown into a fountain.

The contradictory aspect of urban life had won the game. Maybe this destructive eruption was a sign of frustration: We had given these people two days of Arcadia, or two days of liveliness, or two days in the 'space of possibilities'. At the end of the day, when all this was taken away, a destructive ritual marked the return of day-to-day, not so exciting normality. The human scale had given in to anonymity, unlimited.

Conclusion

In the search for multi-ethnic cultures in an urban context, public space is the core where societal change can take place. Public space in central European cities is about to lose its meaning. Immigrants, however, bring along their own idiosyncratic traditions of using public space and contribute to a new, open and diverse use that makes European cities livelier.

New 'social know-how', or new routes of contact between the old and new inhabitants, usually occur in out-of-sight places like the tiny green parks next to building blocks (the Viennese call them ironically *Beserlpark*, park for brooms). Professional festivals like the Gay Parade in many European metropolises are one-day events that attract public interest to minority causes, whereas the day-to-day existence of multi-ethnicity still needs a place to be expressed. The Vienna city administration often categorizes differences and assigns specific spaces for every specific group of city-dwellers. This method avoids ethnic conflicts, but does not solve them. The multi-ethnic festival, however, is a moment of encounter and 'social know-how' on the stage that people call life and urbanists call public space, the latter mainly shaped by people and their social interactions.

In the case of the Waterworld Festival – a platform for social initiatives and economic enterprise – a big, urban and underused space was appropriated for the people who were living in the neighbourhood. This vacant space became a 'space of possibilities'. It invited and empowered people to shape their own surroundings. Multi-ethnic encounters occurred in an open way, with a lot of guests, but also with performances and information booths run by a mix of professionals and local people.

The 15th District in Vienna is a *Zwischenstadt* (Sieverts 1997), an 'in-between city', that is in search of its own culture: it is neither a great city with boulevards, nor a suburb prone to isolation. It is a very specific urban situation, characterized by ethnic and social diversity within those ethnic groups. The future for multi-ethnic use of public space will not be in the form of folkloric, museum-like cultural displays. It will be a hybrid one, where diversity is welcome, where the foreign and the stranger will have a place of their own. Out of this diversity, the public space will develop as a stage for different uses. Some of the inhabitants might choose simply to take a *bain de foule* (a bath in the crowd), feeling free to dress and behave in the way their own cultural patterns prescribe. Some citizens may be encouraged by such an event to choose a more active role.

Notes

1 See the webpage http://oikodrom.com.
2 In contrast to English, the German words for 'you' and 'yours' make a distinction between singular and plural forms.
3 The details and explanations of the variations in the attire of the Turkish female immigrants were provided by Aygen Erdentug over the course of an e-mail exchange on the subject.
4 The term *Tschuschen* is a pejorative Viennese word for 'foreigner'.

References

Boal, Augusto (1995) *The Rainbow of Desire: The Boal method of theatre and therapy*, London and New York: Routledge.
Dumreicher, Heidi (1995) *Bericht über die 'European Conference of Sustainable Cities and Towns' vom 24.–27. Mai 1994 in Aalborg, Dänemark* (Report on the 'European Conference … ' of 24–27 May 1994 in Aalborg, Denmark), Vienna: Magistrat der Stadt Wien.

Dumreicher, Heidi and Kolb, Bettina (2001) 'Sieben Thesen zur Qualität der Stadt' (Seven theses on town and city quality), *Oikodrom Stadtpläne* 24: 7–44.
Dumreicher, Heidi and Levine, Richard S. (1996) *Stadthügel Wien Westbahnhof* (Hilltown Vienna Westbahnhof), Vienna: Projektbericht.
Dumreicher, Heidi and Levine, Richard S. (1999) *Ein urbanes Nachhaltigkeitsprojekt* (An Urban Sustainability Project), Vienna: Edition Dumreicher.
Dumreicher, Heidi, Levine, Richard S., Yanarella, Ernest J. and Radmard, Taghi (2000) 'Generating models of urban sustainability: Vienna's Westbahnhof sustainable hilltown', in Katie Williams, Elizabeth Burton and Mike Jenks (eds), *Achieving Sustainable Urban Form*, London: E & F.N. Spon.
Levine, Richard S., Yanarella, Ernest J. and Radmard, Taghi (1998) 'Sustainable hilltown Vienna: a comprehensive approach to the future of the European city', in *Proceedings of the Conference 'Rebuild the European City of Tomorrow', Florence, Italy: 1–3 April 1998*, Florence: ETA.
Ritzer, Georg (2000) 'Die Gesellschaft als organisierte Erwartungs-Enttäuschungs-Spirale' (Society as an organized spiral of expectation and disappointment), in Uwe Schimanek and Ute Volkmann (eds), *Soziologische Gegenwartsdiagnose*, vol. 1, Stuttgart: UTB.
Schütze, Jochen K. (2000) *Vom Fremden* (On Strangers), Vienna: Passagen.
Sennett, Richard (1994) *Flesh and Stone: The body and the city in Western civilization*, London: W.W. Norton.
Sieverts, Thomas (1997) *Zwischenstadt: Zwischen Ort und Welt, Raum und Zeit, Stadt und Land* (The In-Between City: Between place and world, space and time, town and countryside), Braunschweig, Wiesbaden: Vieweg.
Stadt als Bühne (2000) 'Stadt als Bühne' (City as Stage), *Oikodrom Stadtpläne* 23: 37–51.

13 Contested urban space
Symbolizing power and identity in the city of Albuquerque, USA

Eveline Dürr

Introduction

Theoretical approaches and ethnographic studies related to the built environment in cities show that the spatial setting is closely connected to the specific ideologies and messages of power-holding individuals or groups. This phenomenon has been addressed by scholars from different disciplines with the conclusion that the design of space constitutes a cultural production that reflects urban hierarchies and shapes social interaction. Geographers point to the textual quality of the urban landscape by defining it as a signifying system that transmits and encodes specific information on a symbolic or metaphorical level (Donald 1992; Duncan 1990; Duncan and Ley 1993). Urban sociologists analyse the forms of occupation and the exclusion of social groups in the public and private spaces of the city (Sassen 1991, 1996; Zukin 1995, 1996). Referring to both the political and the symbolic economy of space, the visual representation of groups at the heart of global cities is detailed, focusing on individuals (such as the homeless or racial minorities) who are not usually considered in the official urban image. The interaction of diverse cultural and ethnic groups within the same spatial environment results in contests and conflict, as each of these groups strives to transmit different messages through the representation of space – messages that may be related to economic, social, cultural or political interests. This holds especially true for what Marc Augé (1992) has called 'anthropological places', meaning places comprising a highly symbolic and identity-creating quality.

In the following case studies from the city of Albuquerque in New Mexico, based on a spatial ethnography framed by historical context, I will offer an anthropological perspective to analyse the struggle between ethnic groups over the control of spatial representation in the traditional centre of the city.[1] The focus is on the cultural strategies used by three different groups (namely, the Hispanics, the Anglo-Americans and the American Indians) to claim, occupy and defend public space that cannot be owned privately.

Historical and present context

In 1598 Juan de Oñate conquered the Indian Pueblos – 'pueblo' being the term used when referring to the Indian settlements or villages in this region – along the Río Grande and claimed the land called 'Nuevo México' (New Mexico) for the Spanish crown and the Catholic Church. It formed the northernmost province of New Spain and was colonized by Spanish settlers. In 1706 a small settlement – consisting of a few houses and a church – was founded on the lower Río Grande by the Spanish conqueror Francisco Cuervo y Valdés. He named it 'Alburquerque', in honour of the thirty-fourth viceroy of New Spain – Francisco Fernandez de la Cueva Enríquez, duque de Alburquerque[2] – who resided in the capital of Mexico. The settlers lived in poor conditions, on a dry and parched land. Their economy was mainly based on stockbreeding and irrigated agriculture. The farmers constantly felt threatened by Indian attackers, who were plundering their fields, stealing the stock and making off with their women and children (Stanley 1963). As a result of the geographic location and harsh living conditions, the province was economically poor and socially isolated. This situation changed only with the independence from Spain in 1821 and the subsequent abandonment of the colonial Spanish law excluding foreign trade. Consequently, merchants from Missouri and the East Coast moved into this region, triggering a stronger Anglo-American influence in the Southwest. After 1821 New Mexico was, politically, a part of the independent Mexican Republic. At this point, all inhabitants, including the Indians and the Mestizos – that is, the offspring of a Spaniard and an Indian – became Mexican citizens and had, officially, equal rights.

After the so-called 'Mexican–American War' in 1846 and the treaty of Guadalupe Hidalgo in 1848, when Mexico lost nearly half of its territory, the new national border of the USA ran along the Río Grande river. As compensation, the USA paid $15 million to Mexico. However, the term 'war' does not exactly define the incidents that took place in New Mexico, since the occupation had been mainly bloodless and the great majority of the inhabitants had not seriously rejected American sovereignty. In several cities the authorities had not even considered fighting the US Army. In Albuquerque, the residents fired guns from their church to welcome the US soldiers and, in the capital Santa Fe, the governor changed the flag – voluntarily and without any resistance – to the American one (Jenkins and Schroeder 1993 [1974]: 47; Johnson 1980: 15). The main reasons for such behaviour were economic advantages, especially the opening of trade routes and the influx of Anglo traders with new commodities from the east, which now flooded the neglected province. In addition, the US Army offered protection against Indian attacks. Furthermore, New Mexico had been under the rule of the Republic of Mexico for only a very short period. Since the Mexican administration had been far away and not really very concerned with this northern and economically weak province, the New Mexicans had never developed a deep loyalty to the Mexican government (Gonzales 1993a). In 1850 New Mexico officially became a US territory. Due to this new political situation, all New Mexicans became US citizens.

The most important event, by the end of the nineteenth century, was the construction of the Atchison, Topeka & Santa Fe Railway, connecting the industrial cities rising on the East Coast with the rural farming states in the south and in the west of the country. The railway triggered economic progress. Trade with stock and lumber was intensified, new enterprises opened up, while hotels and restaurants provided further employment. Consequently, the migration of Anglo-Americans and Mexican nationals, in search of jobs along the railway line and in the new businesses, to New Mexico increased. Besides the economic development, the geographic location of the city (1,500 metres above sea level) also contributed to its growth. Due to the excellent quality of the air in the area, patients with lung disorders sought recovery in Albuquerque. Sanatoriums and health centres were founded, stimulating the development of the medical sector. After medical treatment, many of the former patients settled, with their families, in the city (Oppenheimer 1962: 57).

The railway did not exactly run through Albuquerque but 1.5 miles to its east. A new district, called 'New Town', began to flourish along the tracks. It soon developed into a commercial and business centre, dominated mainly by Anglo-American professionals. So Albuquerque itself, now called 'Old Town', became isolated and neglected. The spatial and ethnic segregation, along with the difference in economic development, in the Old Town and the New Town deepened the social gap between them. To be of Anglo-American descent became associated with progress, prosperity and education, while the Hispanic and Indian populations were considered backward, poor and simple-minded. Due to their descent and language, rising to a higher social class was mostly denied to the latter group and reserved almost exclusively for the Anglo-Americans.

In the 1930s the promotion of tourism steadily emerged. When the transportation system was improved, the new infrastructure facilitated comfortable journeys to the Southwest. The ethnic diversity of the population in New Mexico and its inherent economic potential was utilized as an essential characteristic of commerce.[3] The Old Town was discovered by artists, bohemians and business personalities who idealized the living conditions in this district. By renovating visual and symbolic spatial designs, they adapted the Old Town to their romantic image of the Southwest. They purchased old buildings and modified them to open art galleries and gift shops. In the new design, special emphasis was given to Spanish culture and the Catholic missions, along with the mystique of the Pueblo culture. At this point, some Pueblo Indians started to offer souvenirs to tourists in the Old Town. They sold local crafts, especially jewellery, as hawkers – called 'vendors' – presenting their goods on blankets. The transformation of the Old Town into a tourist site triggered a flourishing economy in this previously neglected district, but had social consequences as well. While the history and the cultures of the Southwest retained their mystique, discrimination still existed among the various ethnic groups. The local population felt socially despised and culturally exploited at the same time. Due to inflated property values and taxes, some of the Hispanic residents were forced to move out of the Old Town.

On the eve of World War II, Albuquerque's population began to increase dramatically. The number of inhabitants doubled between 1940 and 1950, climbing from 35,449 to 96,815 (*Albuquerque Databook* 1992: 11). Thousands of people were drawn to the city because of the developing armaments industry, the Air Force and the founding of the Sandia National Laboratories. In 1949 the New Town, now called 'Albuquerque', annexed the Old Town, even though this was opposed by the majority of the residents of the Old Town. By the end of the 1950s a new Historic Zoning Ordinance was passed by the city council to control the architecture-related activities in the Old Town. The new ordinance was applied to the Historic District comprising the Spanish plaza, the church and the adjacent streets (see Maps 13.1 and 13.2). Even though many local residents disagreed with the city plans concerning historic preservation, a commission was established to supervise the future image of the historical centre.

Today, Albuquerque is the biggest city in New Mexico, with a population of nearly half a million. The majority of the residents (78 per cent) belong to the racially defined statistical category of 'white', including white 'Hispanics'.

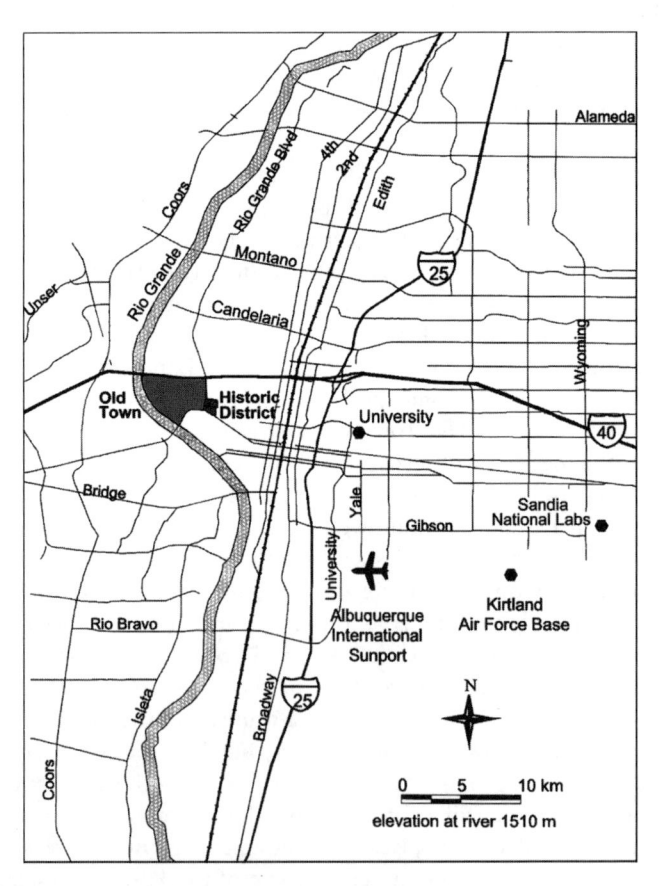

Map 13.1 The city of Albuquerque

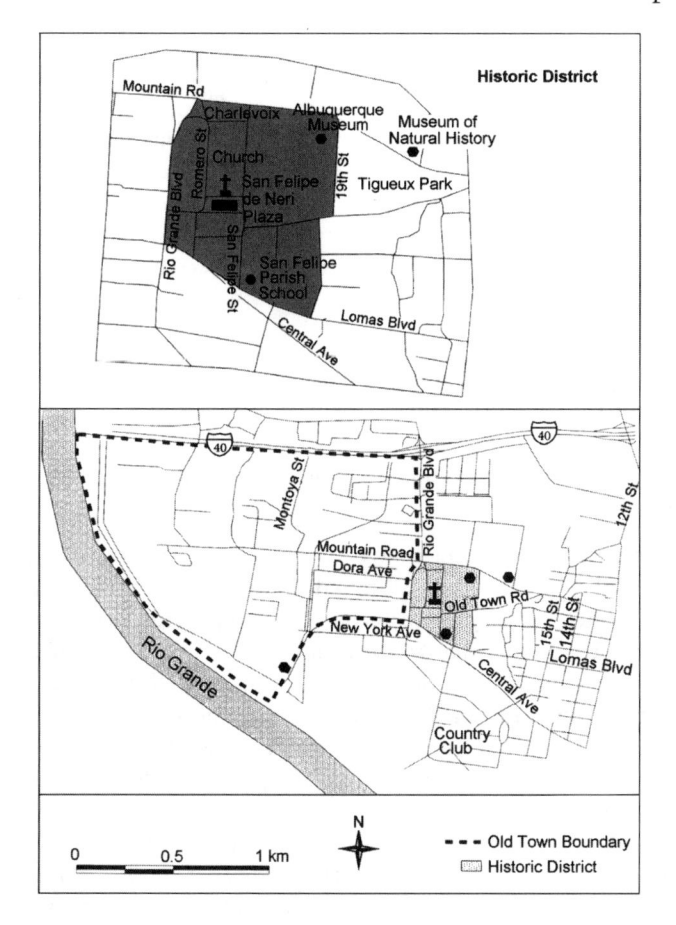

Map 13.2 The Old Town and the historic district of Albuquerque

'American Indians', as well as 'blacks', represent 3 per cent and the 'Asians' 2 per cent of the population. Individuals that could not be classified or denied classification have been grouped into a category labelled 'other race' that represents 14 per cent of the population (*Census of Population* 1993: 13).

In Albuquerque, 34 per cent of the total population of the city is defined as 'Hispanic'[4] – a disputable term. In contrast to the primarily racial terms of 'Anglo-American', 'Indian', 'black' or 'Asian', 'Hispanic' refers to a cultural category that includes individuals of any race relating to those from very different cultural and national backgrounds, ranging from Latin America to Spain.

The presence of the Indians, mostly the Pueblo and the Navajo, in the urban public is an important feature of the image of the city. While only 3 per cent reside in Albuquerque, many Indians commute back and forth between the city and the surrounding reservations. They feel socially and culturally more connected to the reservations but participate in city life, working in different sectors, using urban facilities and institutions like schools, hospitals and so on (Prindeville 1992: 24).

In the official representation of the city, the various ethnic and cultural groups are portrayed as living together in peaceful unity, which does not, however, correspond to reality. Tensions and sociopolitical problems have continued to mark the relationship between them from the period of first contact until the present – a matter that is partly reflected in the contest for the spatial design of the historical city centre.

Representing and contesting space as struggle for power and control: two case studies

Case study 1: Instrumentalization of cultural stereotypes to protect space from competitors

At first sight, the actual depiction of the Historic District of Albuquerque reminds one of a theme park. Scattered around the Spanish plaza and the church, the owners of the approximately 200 shops and over 30 art galleries strive to present a picturesque and romantic image of a historical multiethnic and multicultural place. The built environment is used to evoke stereotype images of the exotic Indian, Spanish and Hispanic cultures. The dominant themes used in the urban landscape are representations of Pueblo traditions, Spanish and Hispanic history, particularly the period of the pioneers and early settlers, as well as the idealized Wild West. The shops advertise and sell local crafts. Moreover, tourists are invited to observe and photograph the 'real' Pueblo Indians, the Navajo and the Zuni, while at work making jewellery. Hence, not only are products advertised and sold but also the Indians themselves are used to attract consumers and are included in the process of commercialization. A similar situation can be observed with the Hispanic culture. The everyday life of the Spanish settlers is portrayed as an adventurous conquest of the 'wild' Southwest. Typical Hispanic crafts like tinwork and religious items are advertised for sale. In this way, 'culture' that is associated with exotic individuals, historical events and material objects becomes a commodity and is offered for sale.

The social characteristics of the tourists in the Old Town range from low-budget backpackers to generous honeymooners. Most of them are travelling through Albuquerque on their way to the more famous tourist sites in New Mexico and in Santa Fe, the capital. The number of visitors to the Old Town is hard to enumerate and can only be estimated. As a clue, one can give the fact that almost 300,000 brochures per year are printed for the tourist centre (*Albuquerque Journal* 6 August 1989: G1).[5] However, the idealization of the Hispanic and Indian cultures does not correspond to sociopolitical and economic reality. In the residential part of the Old Town, almost 75 per cent of the population is of Hispanic descent, in contrast to being 34 per cent of the population of the city.[6] In the Historical District, however, more than two-thirds of the property owners are Anglo-Americans, while one-third is of Hispanic origin.[7] The businesses are mainly operated by the Anglos, with only three shops run by Indians; around fifty tribe members, called 'vendors', sell crafts and jewellery in front of a restaurant on the east side of the plaza.

The diverse groups meet and interact on a daily basis in this quarter of the city. For each group this space has a different function and is associated with a specific meaning. Due to the differences in perception, conflicts related to the spatial representation arise between the various groups. The Anglo shop owners – without any cultural or historical roots in the city centre – perceive the Historic District primarily as an economic resource conditioned by tourism and the invention of a consumer-oriented design. In contrast, the Hispanics consider this space as 'homeland' because it is the area where their community was originally founded. The Indians selling crafts on blankets perceive the same space as a place to maintain and perpetuate their traditions – especially those concerning the production of typical crafts – and as a place of material income. For decades these facts have given rise to controversy over the spatial representation between the competing groups.

Since the emergence of tourism in the 1930s, the vendors had been mostly Indians from the nearby Santo Domingo Pueblo, travelling daily to the Old Town to sell their crafts to tourists. However, in the 1970s, individuals of Anglo-American descent joined them, also offering souvenirs on blankets. This led to rivalry with both the Pueblo Indian vendors and the business personalities in the Historic District. The shopkeepers considered the newcomers to be 'Hippies', disturbing the romantic picture of the Old Southwest. In their view, the 'Hippies' created an atmosphere more like a flea market and did not fit into the historic scene. This opinion was shared by the Indian vendors, who were not very pleased with the new competitors. However, the young Anglo vendors opposed this rejection and resented the term 'Hippie', considering themselves to be young business professionals (*Albuquerque Journal* 9 May 1976: A1). The shopkeepers asked the city council of Albuquerque for support with regard to this matter. As a consequence, the city council drafted an ordinance that would allow the Indians from the Santo Domingo Pueblo exclusively to sell crafts in a restricted area of the Historic District – that is, only on the sidewalk in front of a restaurant to the east of the plaza (*Albuquerque Journal* 11 May 1976: A6). But this proposal was challenged by non-Indian vendors, as well as by members of other Indian tribes. Finally, after intense debates (including the consultation of experts in city planning), the city council passed, in 1989, an ordinance stating that only Indians who are members of federally recognized tribes in New Mexico and Arizona would be granted permission to sell handmade, traditional-style jewellery and handicrafts. No machine-tooled craft would be allowed. Every two years, the forty-five different vendors to receive such a licence were to be selected by a lottery system. A fee of $100 was required for the licence (*Albuquerque Journal* 25 January 1989: D1).

The ordinance was approved in 1989, following an existing example in the state capital of Santa Fe, where tribe members only are permitted to sell souvenirs in front of a historical museum. In this case, they are considered to 'belong' to the museum as a 'living exhibit'. In addition, the city planners argued that the cultural and historical integrity of an area should and could be legally protected, like the French Quarter in New Orleans. But in Albuquerque, the situation is different. In contrast to Santa Fe, the vendors in Albuquerque do not sell in front of a museum but under a portal in front of a restaurant. Therefore, it is doubtful whether they can be included in the programme of a museum, in terms of being considered as the 'living objects' of an exhibit.

With this development, the Anglo-American shopkeepers, as well as the city planners who regarded the Indians as part of the touristic image of the Old Town, pretended to be concerned about the loss of the Indian tradition by allowing others to copy it. In the newspaper they stated that:

> selling jewelry from behind a blanket is a Native American tradition. ... If you and I were to sit down behind a blanket and sell jewelry it wouldn't be the same because that's not what our culture normally does.
> (*Albuquerque Tribune* 13 June 1989: D1)

Selling merchandise on blankets was considered to be part of the Indian heritage and culture that should be protected against the increasing influence of the Western world. Therefore, only Indians could claim traditional rights to sell crafts on blankets in the Old Town. Individuals, especially those of Anglo-American descent, should not be allowed to copy this aspect of Indian culture. Since the Indians have been selling jewellery at the plaza for at least four decades, they can claim traditional rights and expect special treatment. This view was argued by a woman from the Santo Domingo Pueblo. She stated that her mother sold crafts in the Old Town and in Santa Fe, and that was why she was eager to carry on this tradition. She insisted that Indian jewellery should be made and sold exclusively by the Indians (*Albuquerque Tribune* 13 June 1989: D1).

Not only 'culture' but also 'tradition' and 'traditional rights' were used in the arguments to defend the exclusivity of the licence, for the Indians, to sell on blankets in the Historic Zone. However, as observed in the local press, these statements led to controversy among the vendors themselves. Some Indians felt abused, as if living in a zoo and being used as bait to lure tourists (*Albuquerque Tribune* 22 May 1989: A3). The vast majority, however, did not share this criticism and did not even think of fighting against cultural exploitation and commercialization. Moreover, they declared publicly that they did not mind being used to attract tourists, nor being referred to as a 'living exhibit' and being treated like exotic objects. Some even stated that they were happy to be a part of it (*Albuquerque Tribune* 13 June 1989: D1). This does not at all correspond to other official Indian political and ideological declarations in the USA, claiming dignity, equal social rights, the right of self-determination and, therefore, strictly prohibiting the commercialization of Indian cultures as commodities. In all of the Indian reservations of New Mexico, for example, sketching and taking pictures is subject to rigid laws and regulations. It is strictly forbidden for non-tribal members to make any profit out of Indian cultures. But, in contrast, in the case of Albuquerque, the Indians agree to serve as a tourist attraction. Why? By declaring that selling products on blankets is part of their 'culture' and 'tradition', the Indians are making use of the common stereotypes about a 'typical Indian' to protect a contested space from competitors.

Along with 'culture' and 'tradition', some social classifications were also integrated into the public debate. Some Pueblos even devalued their work, in the local press, as a business for the uneducated people. A woman from the Santo Domingo Pueblo commented in the newspaper:

> They [the non-Indian vendors] have more education and should work – not sit down there. ... I would be ashamed if I was white selling down there. I am not going to leave Old Town. I want to sit down there all my life.
> (*Albuquerque Tribune* 8 August 1989: A3)

Another public statement referred to the better education of non-Indian individuals who would be able to find another type of employment, other than that available to the Indians (*Albuquerque Journal* 13 July 1989: E1; *Albuquerque Journal* 8 August 1989: D1). By relating to a low level of education, social inequality and prejudice are used to exclude non-Indian vendors from the contested space. Those of Anglo-American descent, especially, are portrayed in the public discourse as being better educated and more sophisticated than the Pueblo Indians or the Navajo who belong to the lowest economic and social class in the matrix of the wider society. This example shows that 'culture', 'tradition' and social stigma can become instrumental in strategies aiming to claim and defend public space in order to increase economic profit and to advance specific interests.

But the ordinance of 1989 was challenged recently. Members of other Indian tribes claimed their constitutional rights to equal protection and opposed the limitation on sales, exclusively allowing New Mexican Indians and the Navajo Nation to offer their crafts in the Historic Zone (*Albuquerque Journal* 16 December 1989: E2). In addition, the Hispanics got involved in this conflict and also opposed the ordinance. They wanted to join the Indians in holding exclusive selling rights, arguing that the core of the Old Town, namely the plaza, 'belonged' historically to their culture because of its Spanish origin. Therefore, they could also claim historical rights on that space – not just the Indians (*Albuquerque Journal* 27 September 1990: D1; *Albuquerque Journal* 9 October 1990: D1). With this development, Albuquerque, a city that took pride in its tri-cultural heritage, risked being sued for racial discrimination.

Finally, in October 1990, the court decided that the exclusion of non-Indian vendors from the Historic Zone was not in keeping with the US

Constitution. For a short period, all of the vendors were banned from hawking on sidewalks in the Historic Zone. But again, there were objections. While some of the Anglo businessmen supported the removal of the street vendors since they viewed them as unfair competitors selling cheaper products – such as imported low-quality crafts, without having to pay taxes or rent for a shop – the others wanted them back. The majority of the shopkeepers came to the conclusion that the Indian vendors contributed to the mystique and unique atmosphere of the Old Town. According to them, the presence of the Indians matched the image tourists expected from New Mexico. Without them, the plaza did, indeed, loose part of its typical charm and colour (*Albuquerque Tribune* 16 January 1991: A9). To emphasize their arguments and increase public pressure, the shopkeepers organized demonstrations and put up signs in their shops supporting the return of the vendors to the Old Town. More than 80 per cent of the 170 merchants spoke in favour of the return of the Indians (*Albuquerque Tribune* 12 March 1991: D1; *Albuquerque Tribune* 1 April 1991: A1). The Indian vendors too had set up signs to attract attention to their case. At this point, along with the various groups in the Historic District, the whole population of the city, as well as the tourists, got involved in this debate. The results of a public survey showed that 91 per cent of the residents of the city voted in favour of the return of the Indians to the Old Town (*Albuquerque Tribune* 1 April 1991: A1). The tourists were asked to sign petitions supporting the vendor issue (*Albuquerque Journal* 13 January 1991: F3).

However, the public controversy on this matter threatened the romantic image of a peaceful tri-cultural city. The city council decided to put an end to the debate. In 1991 hawking in the Old Town was permited again, without any racial restrictions. The number of permits was limited to forty-five per year, at a cost of $125 each. The products were to be hand-crafted goods, made by the vendors themselves or by one of their family unit. Imported crafts were officially excluded from sale to protect the shopkeepers from unfair competition. A commission was established to control the sale under these restrictions (*Albuquerque Tribune* 21 May 1991: A1). So at least officially, if not in practice, the clearly defined city regulations have put an end to this controversy.

Today, the vast majority of the vendors in the Old Town are members of Indian tribes; only a few are still non-Indians. The reason might be the fact that the height of the 'Hippie fashion' – the issue that originally triggered the spatial control debate – is over. Yet resentment exists among the various groups. The Indian vendors consider non-Indians to be intruders claiming 'their' space and copying their cultural traditions. Some Anglo shopkeepers have doubts about the vendors selling only hand-made crafts and see them as unfair competition, selling cheap imported products of lower quality. On the other hand, the Hispanic residents did not get deeply involved in this public controversy since they were not as much concerned with the tourist business as the shopkeepers.

Case study 2: Instrumentalization of history and tradition to claim spatial rights

Although the Historic District resembles an outdoor museum, there is an important difference. Centrally located at the north side of the plaza, the San Felipe de Neri Church visually dominates the scene. Since its foundation in 1706, services have been held without any interruption and many church members are proud of the fact that their lineage dates back to this time. Today, the Catholic parish has an active membership of roughly 1,000 families who are mostly of Hispanic descent.

The promotion of tourism and the increasing Anglo-American influence in the Old Town since the 1930s, provoked modifications to the self-definition and identity of the local Hispanics. They developed a variety of cultural strategies as a reaction to racism and discrimination. In terms of redefining identity, special emphasis was given to the European-related lineage and the Spanish cultural heritage. By stating that they were the descendants of the people who first conquered and originally colonized the land – the Pilgrims arrived after the Spaniards in the New World – they claimed the 'original right' to control it. Focusing on European bloodlines also distinguished them from other groups that possess Spanish surnames but are either of Mexican origin or are Latin American nationals. In addition, belonging to the white and dominant part of the wider society is documented with the notion of being 'Spanish'. The real blood relationship to the pioneers is proved by the relatives living in Spain, their original homeland.

Besides racial categories, genealogy and historical events, many Hispanics take pride in the cultural heritage of the *conquistadores* as demonstrating the difference between themselves and others. To emphasize cultural differences with Anglo-Americans, it is usually pointed out that the Pilgrims were poor religious refugees who feared for their lives in their homeland, whereas the Spaniards arrived proudly from a well-established country that was, at the time of the conquest, one of the most powerful and flourishing nations in the world. The exceptional nature of the Spanish culture is given importance by referring to the so-called 'Golden Age' in Spain during the time of the conquest, represented by names like Cervantes or Lope de Vega.

Another crucial aspect of the Hispanic cultural identity is the Catholic faith. Religion was one justification for the conquest and attempt – in colonial terms – to civilize the 'savages'. Even today, individual as well as community life are deeply embedded in the Church and in Catholic rituals. Catholicism and Catholic education constitute basic identity-creating elements, shaping social norms and values. A parish member expressed the impact of historical events and cultural patterns on the self-image of the Hispanic community in her own words:

> So our grandparents' parents were born under the Spanish flag. ... all of this territory, was *Nueva España*. It was New Spain. It was all controlled by Spain. We were brought up being Spanish. We spoke Spanish, we were

Catholics ... . We had our own sense of who we are and we weren't gringos. We weren't Americans. And so we retained our tremendous pride in being Spanish and the pride was not in a nation that we didn't know, cause we don't know necessarily, but in our language, our music, our customs and traditions and everything. And we're just very proud of that. Then, when the Americans came, they took our land, they thought of us as being less than equal. We were cheap labor. ... Where we tried to say, you know, we're descendants of royalty, you know, don't misconstrue that statement, but that statement means that we are somebody.[8]

By glorifying the past, honouring the Spanish colonists and paying tribute to the Catholic Church, cultural 'otherness' and European descent are expressed in the urban public.

The Hispanics endeavour to represent these distinguishing features not only in sociopolitical discourses but also in the spatial environment of the city. In contrast to the commercial representation and the primarily economic function of the Historic District with regard to the Anglo-American and Indian groups, the local Hispanics perceive this space as 'homeland'. In their view, it is the place where Albuquerque was originally founded by the Spanish pioneers, whom they consider to be their forefathers. Because of the history and location of the church, this space comprises an important sacred aspect as well. Church ceremonies and processions are conducted on the plaza and in the nearby streets, and services are held on a regular daily basis.

The various perceptions and functions of the same space by these interacting groups leads to competition and conflict over its design. The analysis of the rising controversies indicates that they do not just relate to the representation of the spatial layout. Rather, they symbolize the struggle for power and equality in this society. This interpretation is highlighted in the following examples.

In the 1980s a public debate arose when a restaurant located at the east of the plaza in the Old Town applied to the city council for permission to serve liquor. This application was heavily opposed by members of the San Felipe de Neri parish. The opponents rejected the endeavours of the proprietor of the restaurant, stating that the residential quality of the area would decrease as commercialization increased. They referred to the New Mexico State Liquor Control Act of 1939, which prohibited the sale of liquor within three hundred feet of a church or school. However, this provision had always been subject to a waiver by the city authorities (Atencio 1985a: 3). In spite of the arguments of its opponents, the restaurant applied to the city council of Albuquerque for a waiver. The application was supported by most of the shopkeepers – as well as by those Hispanic parish members who were involved in the tourist industry. The Hispanic supporters were considered traitors by their opponents and called 'potatos' – that is, 'brown' on the outside, 'white' on the inside. In order to emphasize their objection, the opponents organized anti-liquor demonstrations; they were joined by other religious communities, including the Anglo Baptists, creating a rather uncommon community in New Mexico. In addition, the archbishop got involved and publicly voiced his support during a mass in San Felipe de Neri Church. As a sign of protest, in case the waiver should be approved by the city council, he authorized the closing of the church when services were not being held (Atencio 1985a: 3). His support of the cause gained state-wide publicity in the media and a controversial debate arose over this matter. The Catholic Church and the clergy were charged with hypocrisy because alcohol is common in most of the Catholic religious ceremonies and festivities, including the Holy Communion. Besides, historically there had always been taverns as well as brothels in the Old Town, in very close proximity to the church. It is also a fact that the Jesuits, who resided almost a hundred years in the San Felipe de Neri Church, were famous for the cultivation of fine wines (Atencio 1985b: 355).

When the city council approved the waiver, this time the opponents accused the city administration of favouring commercialism and not protecting the social and cultural integrity of the historical centre. What is remarkable is that the criticism was not directed against the liquor licence but against being manipulated by 'outsiders'. In other words, the main reason for opposing the waiver was not the rejection of alcohol but the struggle over the question of who dominated and who directed how the historical centre of the city was to take shape. It this sense, this incident can be considered as a public contest revealing the power structure in the Old Town. These aspects are explicitly mentioned in an official statement made by a Hispanic parish member:

> It's no longer a question over just a liquor license. The issue now is when do you stop letting outsiders tell you what to do. *Ya basta*. ... Our ancestors lived here, worked here, suffered here, but it's a heritage many merchants are insensitive to. ... City officials will tell parishioners to seek permission from local merchants before staging religious activities, such as processions. ... Every time we want to do something, we have to beg the outsider, the merchants and the city, and that's what makes us so angry.
> (*Albuquerque Journal* 24 July 1980: A18)

In spite of the harsh arguments that took place in the urban public, the restaurant won this battle. The waiver was approved by the city council of Albuquerque with a slim majority: a vote of five to four.[9] But this defeat concerning the use of space triggered another protest. As a sign of resistance, the church – being the main symbol of the traditional Hispanic community – was closed to the public. Its entrance was draped with black funeral hangings and its bells were tolled. By doing so, the church went into mourning (Atencio 1985b: 394). This was the first time in almost three centuries that the church was closed to the non-worshipping public. To some parish members this event marked the end of their community life and symbolized the death of the traditional neighbourhood. A large sign on the entrance explained the reason for the closure of the church.

The opponents continued their battle by filing suit against the City of Albuquerque, which delayed the issue of the liquor licence for two years

(Atencio 1985a: 5). But finally they had to accept their losses in this contest. To this day, some of the former activists refuse to set foot in the restaurant that was involved, considering it a site of defeat and humiliation. One of them told me: 'I have not gone into *La Placita* [the name of the restaurant], nor will I go till – not even till the day that I die.'[10]

Today domination over the Old Town becomes controversial when traditional rituals are performed: when, for example, church services are held while noisy gunfights for tourists take place in the streets. Another disputed event is the festival honouring the patron saint San Felipe de Neri which is held on the first weekend in June. The parishioners conduct a procession through the streets and organize a party. Food and soft drinks are sold around the plaza, with booths and musicians entertaining the crowd. During this celebration, the streets are closed to traffic, leading to conflicts with the shopkeepers who complain about financial losses. The parish members interpret these complaints as an attack on their community life and as typical of the alien intruders, trying to control 'their' space without paying respect to local traditions. On a symbolic level, as well as in physical terms, the Hispanics reoccupy and reconquer the plaza during this event, displaying crucial features of their cultural identity. Changing the meaning of the urban environment from an economic to a sacred and social place, even for a short period, reveals their cultural roots and religious traditions. In this example, public space is used as a stage to carry out conflicts that must be contextualized and interpreted within a historical frame.

Conclusion

Ethnic and social groups develop specific cultural strategies to occupy, control and defend public space. The strategies chosen relate to the perception and function of the urban environment by the relevant group. Diverse interpretations of the same space lead to disputes over its control and representation. Urban conflicts are rooted in the complex interactions of the rearticulation of ethnic boundaries, the mobilization of cultural heritage and the assertions of power and hierarchy. The struggle over the design of a city area reflects the impact of space on identity patterns and inter-ethnic relationships.

In the historical centre of Albuquerque, the cultures of the Southwest are portrayed in a romantic way. However, behind the image of a peaceful co-existence of different ethnic groups is a struggle for control over spatial representation and the messages conveyed. The shopkeepers, mainly of Anglo-American descent, perceive the Historic District as an economic resource, primarily created to attract tourists. By commercializing and converting it into a tourist-oriented theme park, they construct an artificial image of the Southwest, including the advertisement of Indian and Spanish traditions, encouraging tourists to buy 'culture'. The Pueblo Indians and the Navajo, who sell their crafts on blankets, make use of the existing cultural stereotypes and social stigma in the public discourse to claim spatial rights and to protect themselves from competitors: for instance, by promoting the selling of crafts on blankets as an Indian cultural element and as the traditional work for uneducated people. Thus cultural stereotypes and social stigma become strategically instrumental in achieving their aims and increasing their profits. This example shows that 'culture' and 'tradition', as well as social categories, become resources to defend public space in order to prevail over competing interest groups. On the other hand, the local Hispanics consider this space to be 'homeland' and therefore claim the 'original right' to control its design, as shown in the case of the liquor dispute. In addition, they endeavour to dominate the spatial setting, using it for self-representation, hence demonstrating historical roots and European-related lineage. They give special emphasis to the Spanish conquest and 'civilization' by performing religious ceremonies and parades to protest the Anglo dominance. The association between identity, cultural meaning, representation of space and social power is embedded in an historical context that shapes present inter-ethnic relationships in Albuquerque.

Acknowledgement

This paper is part of an investigation which was made possible by a grant from the Deutsche Forschungsgemeinschaft, Bonn, Germany.

Notes

1. This research is based on the sources in various archives and on fieldwork conducted in the city of Albuquerque, New Mexico, during a nine-month period in 1997 and 1998. Important sources for both case studies were the archives of the local newspapers, as well as the studies of Atencio (1985a, 1985b) and the articles of Gonzales (1993a, 1993b).
2. With the beginning of the American period in 1846, the 'r' was dropped and the city was renamed 'Albuquerque'. For more details on the name and the foundation of Albuquerque, see Chávez (1972: 5) and Dürr (2000: 82ff.).
3. The development of the tourism sector with an artificial image of the US Southwest and its consequences for inter-ethnic relationships is documented in several studies. Wilson (1997) showed that the image of the state capital of New Mexico, Santa Fe, was based on the creation and celebration of a myth, invented by Anglo-American newcomers about the turn of the century. Rodríguez (1989: 77–99) sheds light on the mystique of the art colony society in Taos, New Mexico, and the modern economy of tourism. She referred to a model of 'ethnic tourism' – focusing on cultural exoticism – as a basic concept for studying the interaction between tourists and ethnic groups.
4. A further discussion of this term, analysing social and political implications, is provided by Gonzales (1993a).
5. It is estimated that three to four million tourists visit the city of Albuquerque per year. A survey conducted in 1996 showed three million visitors, although this figure is not accurate because the visitors were counted daily which meant that people could be counted twice (information by telephone, Department of Tourism, Santa Fe, New Mexico, 3 March 1998). The average travelling party of 2.5 people who visit Albuquerque spend an average of $159 a day (*Albuquerque Tribune* 11 December 1989: C4).
6. This data is based on the census tracts 25 and 26, including the Old Town and parts of adjacent neighbourhoods (*Albuquerque Databook* 1992: 3).

7 Personal communication of the board members of the Historic Old Town Property Owners Association, September 1997.
8 The interview was conducted by the author on 11 November 1997. To guarantee anonymity, I have avoided mentioning informants' names throughout this article.
9 According to Atencio (1985a: 4) three Hispano councillors and one Anglo voted against the waiver, the other five decided to support the restaurant's application.
10 The interview was conducted on 4 September 1997.

References

Albuquerque Databook (1992) *Albuquerque Databook*, Albuquerque: City of Albuquerque Planning Department.
Atencio, Tomás (1985a) *The Old Town Liquor Dispute: Social change and conflict in New Mexico* (Working Paper 112), Albuquerque: Southwest Hispanic Research Institute, University of New Mexico.
—— (1985b) 'Social change and community conflict in old Albuquerque, New Mexico', unpublished Ph.D. thesis, University of New Mexico, Albuquerque.
Augé, Marc (1992) *Non-lieux: Introduction à une anthropologie de la surmodernité*, Paris: Seuil.
Census of Population (1993) *1990 Census of Population: Social and economic characteristics, New Mexico*, Washington: US Department of Commerce, Economics and Statistics Administration, Bureau of the Census.
Chávez, Fray Angélico (1972) 'San Felipe Neri de Albuquerque', in Robert Sánchez *et al.* (eds) *From the Beginning: A historical survey commemorating the solemn rededication of San Felipe de Neri Church, 1706–1972, Old Town Plaza*, Albuquerque: San Felipe de Neri Church.
Donald, James (1992) 'Metropolis: the city as text', in R. Bocock and K. Thompson (eds), *Social and Cultural Forms of Modernity*, Cambridge: Polity Press.
Duncan, James S. (1990) *The City as Text: The politics of landscape interpretation in the Kandyan Kingdom*, Cambridge: Cambridge University Press.
Duncan, James S. and Ley, David (eds) (1993) *Place, Culture, Representation*, London: Routledge.
Dürr, Eveline (2000) *Verortung und Repräsentation von Identitäten in urbanem Raum: Hispanics im Südwesten der USA* (Localizing and Representing Identities in Urban Space: Hispanics in the US Southwest), unpublished Habilitation thesis, University of Freiburg.
Gonzales, Phillip B. (1993a) 'The political construction of Latino nomenclatures in twentieth-century New Mexico', *Journal of the Southwest* 35(2): 138–85.
—— (1993b) 'Historical poverty, restructuring effects, and integrative ties: Mexican American neighborhoods in a peripheral Sunbelt economy', in Joan W. Moore and Raquel Pinderhughes (eds), *In the Barrios: Latinos and the underclass debate*, New York: Sage.
Jenkins, Myra Ellen and Schroeder, Albert H. (1993, 1st edn 1974) *A Brief History of New Mexico*, Albuquerque: University of New Mexico Press.
Johnson, Byron A. (1980) *Old Town, Albuquerque, New Mexico: A guide to its history and architecture*, Albuquerque: City of Albuquerque and Albuquerque Museum.
Oppenheimer, Alan J. (1962) *The Historical Background of Albuquerque, New Mexico*, Albuquerque: Report for the Planning Department.

Prindeville, Diane-Michele (1992) *The State of Ethnic and Race Relations in Albuquerque, New Mexico*, report submitted to Project Change Lévi-Strauss Foundation, Department of Political Science, University of New Mexico, Albuquerque.

Rodríguez, Sylvia (1989) 'Art, tourism, and race relations in Taos: toward a sociology of the art colony', *Journal of Anthropological Research* 45(1): 77–99.

Sassen, Saskia (1991) *The Global City: New York, London, Tokyo*, Princeton: Princeton University Press.

—— (1996) 'Rebuilding the global city: economy, ethnicity, and space', in Anthony D. King (ed.), *Re-presenting the City: Ethnicity, capital and culture in the twenty-first century metropolis*, London: Macmillan Press.

Stanley, Francis (1963) *The Duke City: The story of Albuquerque, New Mexico, 1706–1958*, Pampa: Self-publication.

Wilson, Chris (1997) *The Myth of Santa Fe: Creating a modern regional tradition*, Albuquerque: University of New Mexico Press.

Zukin, Sharon (1995) *The Cultures of Cities*, Cambridge, MA: Blackwell.

—— (1996) 'Space and symbols in an age of decline', in Anthony D. King (ed.), *Re-presenting the City: Ethnicity, capital and culture in the twenty-first century metropolis*, London: Macmillian Press.

14 Conclusion

Freek Colombijn and Aygen Erdentug

Throughout this volume, the two leading questions had been: How does urban space structure the life of ethnic groups? How does ethnic diversity contribute to creating an urban shape or give shape to a particular space? In spite of the fact that we do not claim to have provided completely satisfying answers to these questions, we feel that we can still pull together the threads that have come out. Hence, this chapter consists of empirical generalizations based on the case studies given previously, supplemented with additional relevant literature. Taking the limited number of cases into account, every conclusion must remain preliminary. The points are presented as hypotheses to be tested rather than as the final word on an issue. Having included cases from countries other than North America and Western Europe, we hope that the range of possible relationships between urban space and ethnicity has been widened enough to stimulate further research.

We start with the overall conclusion that ethnic diversity and urban space *do* mutually structure each other, in spite of this relationship being contingent, multidimensional and shifting. Having stated this fact we will pass on to discuss the conclusions that can be drawn from the more detailed questions presented in the introduction, such as: To what extent do different ethnic groups live segregated? Why do they insist on living segregated as opposed to being integrated into the cultural mainstream? To what extent are segregation and integration desirable goals? If ethnic groups live segregated, where do they meet? Why do they choose to meet there? Which sections of the town or city do they claim exclusively for themselves? Which regions are forbidden, if not dangerous, zones for other ethnic groups? We have preferred to present the answers, limited to the cases studied, to these questions by grouping them under the following subheadings: residential segregation; ethnicity and class; the appropriation or usage of public space; state intervention in the allocation of space; the fluidity of ethnic identity; and ethnic design in the cityscape.

Residential segregation

One productive way to look at the relationship between ethnicity and urban space is by associating 'the quality of the inter-ethnic relations' with 'the degree of residential segregation of ethnic groups'.[1] Residential segregation is, by the way, not the only way to measure an ethnic group's assimilation into wider society. Another key measure is intermarriage. But residential patterns often indicate larger patterns of social mobility (Phillips 1996: 435). The fact that various ethnic groups live segregated is, of course, not a new insight. It was a cornerstone of the Chicago School which led to successive similar studies. However, the chapters in this volume not only show that the relationship between ethnicity and residence is valid in all parts of the world, but that it continues to play an important part into the twenty-first century, taking on many forms.

The initial point to be made in this respect is that neither ethnic (or racial) segregation nor the quality of inter-ethnic relations are static and a 'state of affairs' situation, but rather they are dynamic processes. Segregated and desegregated existence for one community can range from total segregation to assimilation, with avoidance, cooperation and integration as the stepping stones. Likewise, the association between ethnic segregation and inter-ethnic relationships is dynamic besides being reciprocal. The chapter on Arab–Jewish relations in the 'Little Triangle' (Chapter 4) is a case in point. As the relation between these two groups became more strained over the years, the degree of residential segregation increased. Conversely, the more the Jewish section was enlarged to occupy a disproportionate share of the land and the more the Arabs were forced to live in confined space due to the circumscribed developments of their towns, the more highly strung the inter-ethnic relationship became. Another excellent example in the same vein is the Philadelphia Chinatown (Chapter 8) that evolved over the course of a century, during which period the Chinese immigrants displaced other people. Nowadays, the 'others', in the form of big development projects, are invading Chinatown, reinforcing the Chinese identity of the people in this ethnic quarter of the city.

However, if an ethnic conflict breaks out within a territory whose communities were formerly integrated, the situation becomes untenable and usually results in 'ethnic cleansing'. This tragic process occurred in Beirut during the civil war among the Christians and Muslims. A similar turn took place during the secessionist activities in the former Yugoslavia. In Mostar, for instance, the ethnic segregation that emerged among its multi-ethnic inhabitants, following a long period of peaceful and integrated co-existence, was complemented with the wilful destruction by Bosnian Croat forces of the famous Old Bridge (Jezernik 1999). The single-arched bridge over the Neretva River had been its most famous landmark, linking the double settlements of Mostar bordering both ends of the bridge, as well as being the symbol of a multi-ethnic Bosnia-Herzegovina.

One reason for the dynamic aspect of ethnic segregation is that migration to cities is an ongoing process. Several chapters in this volume show that the order of arrival of immigrants has a big impact on both their social position and on the spatial niche they eventually come to occupy. This is observed clearly in Shahshahani's analysis (Chapter 10) of the complex composition of a district in

Tehran where the first settlers have been honoured by toponyms – that is, with place names derived from their family names. In Vienna the position of recent immigrants is made explicit by the term 'New Viennese', as opposed to 'Old Viennese' which is used to denote former inhabitants. The fact that there is a pecking order among the wave of immigrants – first come, first served and established – is nowhere clearer than in Toronto. There, the white population, once colonists themselves, has acquired a hegemonic position *vis-à-vis* later arrivals – namely, the coloured immigrants from Africa and Asia.

Joe Darden zealously maintains (Chapter 2) that residential segregation is the result of ethnic tensions. White supremacy, Darden argues, is an ideology which holds that in relations involving coloured people, the white race must have the superior position; this ideology restricts the opportunities of visible non-white minorities to attain equal neighbourhood qualities. The spatial consequence of white racism in Toronto is residential segregation. When the subordinate 'races' want to move up socially, moving out spatially from the formerly segregated neighbourhoods becomes a necessity. However, this is not at all easy to accomplish. In the comparable case of the USA, in spite of the fact that the law no longer supports racial segregation, the situation concerning the residential segregation of urban African-Americans has changed little since the 1950s. White people made up 94 per cent of the suburban population in 1977 and in 1990 this figure was still 94 per cent. African-Americans remain highly concentrated in inner cities, even when compared with Asian-Americans and Latinos. Scholars generally estimate that economic factors (differential housing costs in suburbs and inner cities, along with the low average income of the African-American population) account for only 10 per cent of African-American residential segregation (Darden 2001: 185–6; Phillips 1996: 437–8). The metaphor of the 'ethnic melting-pot' has remained illusory, since white people have adopted subtle ways of keeping African-Americans out of their neighbourhoods. The tricks employed include race baiting, rezoning, bank redlining (refusal to grant loans in areas around which red lines have been drawn) and racial steering by real-estate brokers (Darden 2001:186; Eriksen 1993: 139–1; McCarthy 1999: 324; Phillips 1996: 438; Quispel 1996: 334–7).

One line of future research to refine on Darden's conclusion that the Vietnamese are excluded from the highest-quality neighbourhoods would be to add a micro-level perspective. Such a perspective might perhaps reveal that an intervening variable can also partially explain his result: The percentage of home ownership – one characteristic of the composite neighbourhood-quality index – is viewed positively by the indigenous white population. Yet recent migrants, a category in which the Vietnamese are disproportionately represented, may prefer neighbourhoods with rented dwellings that can be occupied and left more easily. In other words, there are multiple perceptions of urban space.

It is perhaps fitting to note here that the term 'race',[2] as it is used here, should be treated as a particular mode of ethnicity (Eriksen 1993: 4–6; Jenkins 1997: 74–84). In a racist notion of society, 'races' are clearly bounded entities, visible by outward signs (such as the colour of skin and other physical features) that take on importance because of alleged hereditary personality traits. In reality, however, as is the case with all ethnic categorization, physical appearance only makes a *difference* because of the fact of it being socially signified as such. Despite the biological basis of the phenotype – the group of observable characteristics, produced by the genotype (basic structure of an organism) interacting with its environment, that make up an individual – the boundaries between 'races' are arbitrary. The arbitrariness of 'racial' categories is demonstrated by the Brazilian concept of *branco* (whites), which includes people, such as Japanese and Koreans, that in Canada and the USA would be seen as 'non-whites'. To take 'race' as a particular mode of ethnicity, with fluid, situational and overlapping boundaries, instead of considering 'race' to be an essential characteristic element of somebody's identity is more than playing with words. When racial categories are treated as the outcome of a process of ethnic labelling, the impact of space on identity becomes clearer. So, in an analysis inspired by Burgess, individuals would be living in Chinatown because of the fact that they are Chinese. We think that it would be more insightful to state that individuals will be considered Chinese if they live in Chinatown, so that in another context individuals would take on and would be ascribed a different identity. What makes 'racial' relations special when compared with other inter-ethnic relations is that, foremost, they are hierarchical in nature. Second, they are comparatively explicit and elaborate. Finally, such relations are more a matter of social categorization (by a hegemonic outside group) than group identification by the ethnic group itself (Jenkins 1996: 19–28; 1997: 53–63). It is because of these three features that Darden correctly insists on ascribing such a negative role to white racial supremacy.

Darden's thesis that ethnic segregation has a negative effect on the segregated group is implicitly modified by Shah (Chapter 3) and Guan (Chapter 8). Shah asserts that religious clustering around prayer houses has a beneficial effect on health. Although there is no absolute concordance between the religious categories used by Shah and the 'racial' categories used by Darden, both are roughly referring to the same groups in the same city. The corollary must be that beneficial religious clustering is the other side of the coin, with the damaging 'racial' segregation being the upper side. The discrepancy between Darden and Shah might be explained by the fact that the former focuses on social aspects and the latter on religion and health. Nevertheless, Shah provides support for Darden's thesis by narrating how the Christian (read 'white') majority in one Toronto neighbourhood opposed the erection of a Hindu temple through sheer demagoguery.

According to Guan (Chapter 8), the economically more successful Chinese have left the Philadelphia Chinatown. Yet it is not clear whether they first left and then attained a level of economic prosperity – as Darden predicts – or, vice versa, that their economic prosperity has been a reason to leave Chinatown and move to a different neighbourhood. For those people remaining in Chinatown, the area provides social support. The local church is the focus of community life and the main source of resistance to outside pressure. Particular campaigns of resistance have been conducted against plans for the demolition of the neighbourhood to make room for urban renewal projects, such as an expressway that would have cut Chinatown into two. Consequently, at considerable extra expense, the municipal government constructed the new expressway underground. The remarkable success of the Chinese resistance can be explained, to a certain extent, by the ironic fact that the city administration did not want to challenge the church in Chinatown, originally a white institution. Guan's findings support the general idea that 'ethnic urban villages' or 'urban enclaves' have a high degree of social interaction and give support to their ethnic members. The collective identity helps to maintain the boundaries with other neighbourhoods (Paddison 2001: 199).

Other studies also support the thesis that Chinese immigrants usually profit from clustering in certain urban quarters for economic reasons. Examples can be taken from Florence, Buenos Aires and, indeed, Toronto (Arbide, Marra and Tavormina 2000; Fong, Luk and Ooka 2000; Lucchesini 1993). In all of these cases, the business concerned is 'ethnic business': either serving the exclusive demands of Chinese customers (like medicinal herbs, hairdressers and so on) or providing popular Chinese goods or services to outsiders (such as food, porcelain and – in keeping with the old stereotype – laundry).[3]

Chinese who are professionals or work in the high-tech sector, by contrast, do not wish to stay in Chinatowns. Chang (Chapter 9) demonstrates that the Taiwanese in Silicon Valley feel that a favourable location for their suburban home transfers their social position from the edge to the centre. The spatial movement is thus inversely related to the social movement. In Los Angeles this has already resulted in the ethnic suburb known as 'Little Taipei' (Paddison 2001: 199).

Ethnicity and class

Ethnicity is not the only and often not the most important factor influencing residential patterns. Class is the other major factor. The more a society is ethnically integrated, the more class becomes the predominant factor of social divisions. It is better to define class here in terms of income and consumption preferences rather than in terms of one's relationship to the means of production. The Marxist division of classes is too blunt to be easily related to the refined distinctions between urban neighbourhoods. Class and ethnicity are of course not independent from each other. Ethnicity can have a considerable impact on opportunities for education and jobs, and thus on income. Conversely, income can be one factor by which to draw a line between ethnic groups, as is shown by the Koreans and Japanese in Brazil, who are defined as *branco* because they are relatively well-to-do like 'white' people.

The association between class and ethnicity is visible in both poor and rich neighbourhoods. The size of the dwelling, the materials used, the quality of design and workmanship, maintenance, the yards and details ('artifacts') are all clues to the economic status and ethnic values of the occupants of a building

(Jacobs 1985: 30–59). Where ethnicity and class coincide, a high degree of social homogeneity (but not necessarily social coherence) occurs within a neighbourhood. Where class and ethnicity do not fully coincide, as is usually the case, a considerable degree of internal antagonsim can exist within a neighbourhood. Yet these statements about the relationship between class and ethnicity are still so general that they border on the meaningless.

The relationship between class, ethnicity and urban space becomes interesting when the concept of 'fear' is introduced. Fear of another class can be felt both by people at the upper end of the socioeconomic hierarchy (that is, those people who have much to lose) and by the people at the bottom end (that is, those people who are most vulnerable). The fears of the people at the bottom end, who stand with their backs to the wall, are most realistic. The fear of the people at the upper end is, however, sociologically most relevant, because the upper and upper-middle class have the means (economic power and political clout) to manage their fears. Fear is strongly reinforced when class divisions are intensified by ethnic divisions. In the latter case, fear has a strong impact on urban form when affluent groups isolate themselves or prefer being shut off from the masses of impoverished fellow citizens (read, 'the mob') by hard physical boundaries such as protective walls, canals and various types of wired fences, as well as other pointed or razor-sharp protective devices. It is also the case when the elite demolishes ethnic slums, under the progressive tag of 'gentrification', transferring their residents to the fringes of the city. Several chapters provide ample evidence of this point.

As narrated by Leisch (Chapter 6), Glodok (the old Chinese quarter in Jakarta) provides its residents with a familiar atmosphere, just like the Philadelphia Chinatown. Once again the local place of worship – in this case a *klenteng*, where a mixture of Buddhism, Confucianism and Taoism is practised – is an important centre for the community. However, any notion of the Chinese territory being a safe zone would be a misconception, as the quarter has been the target of anti-Chinese pogroms, beginning with the Chinese massacre of 1740. The latest manifestation of mass anti-Chinese violence in Indonesia happened in May 1998. Chinese women were first raped then killed; shops in Glodok were looted and burnt. So security has become an important factor influencing the Chinese mental map of Jakarta and probably the main reason why those families who can afford to, prefer to live in the new towns mushrooming on the fringes of Jakarta, away from the centre of discriminatory criminal activities.

Similarly, the residential sector of the Hsinchu Science-based Industrial Park in Taiwan, described by Chang (Chapter 9), resembles the new towns in Jakarta in the way that it is a 'gated community' – that is, it is separated from its surroundings by high walls and guarded gates. The inhabitants of nearby villages are not allowed to enter this area. Moreover, the local people who used to live on that spot were forced to evacuate it so that this new residential area could be built. Also in Brazil, the combination of fear of the ethnically defined lower classes and the desire for a privileged, secluded lifestyle of their own determined the choice of the Brazilian elite in locking itself up behind barbed-wire fences or walls (Chapter 7). This has especially been the case when neighbouring a *favela* (shanty town), with the residents of such *favelas* expressing annoyance at being 'walled-in' since they have to make detours around the protective obstacle to commute to other zones of the city. There is a very good chance that a vengeful invasion of elite condominiums, as has been the case in Jakarta, might occur in the Brazilian megacities. The Brazilian notion that the street is a place of menace, where unpredictable disturbances take place, is also found in the completely different culture of Indonesia (Kusno 2000: 117). However, a resentful incursion is not conceivable for the Hsinchu Science-based Industrial Park in Taiwan, probably because the cultural difference between the 'Parkers' (those working and resident in this Industrial Park) and the surrounding villagers is more in terms of class than ethnicity.

Fear has also played a role in the USA. Mike Davis argues that fear has been a decisive factor in the ecology of Los Angeles. New spatial enclaves have emerged, called 'social control districts' (where noxious problems can be scrutinized), 'containment districts' (where social problems are quarantined) and 'enhancement districts' (where social problems are banned). The surveillance adds up to a militarization of the urban landscape (Davis, quoted in Phillips 1996: 429–30). The surveillance and imposed order are not only a problem for the underclass that is kept under control, but also a burden for the upper class. Richard Sennett has argued that affluent white adolescents especially must escape from their protected, 'purified' suburbs to the disorder of the city, in order to develop fully into adults (Sennett 1973 [1970]).

In contrast to the elite territories in Jakarta, the Hsinchu Science-based Industrial Park, Rio de Janeiro and São Paulo – which have been created for the exclusive use of the elite but also display ethnic demarcations – the new luxurious housing blocks in Beirut serve to bring together the various ethnic elites of the city. Although these blocks are far beyond the financial means of the ordinary people, to a certain degree they represent the beginnings of ethnic integration. The case of Beirut also suggests that the elite possibly tend to mingle more easily than commoners do, probably because they share a latent fear of the lower classes.

A very important finding was that the relationship, mediated by fear, between class and ethnicity, on the one hand, and spacial segregation, on the other, is not inevitable. An excellent example of how people of various classes and ethnic backgrounds can mingle is provided by Shahshahani (Chapter 10). She has described a Tehran neighbourhood of just a few blocks, where many groups live together. The contacts are conditional. The places and occasions where the members of different ethnic groups meet are restricted, but women and young children cross the ethnic borders more easily than adult men.

A final note on residential segregation and class is on an important variable that influences its social effects – that is, the question of whether segregation has been enforced or self-imposed. In the cases of black ghettos in Toronto, the Arab areas of the 'Little Triangle' and the *favelas* of Rio de Janeiro, the segregation has been enforced by another – hegemonic – ethnic group. Under such circumstances, segregation becomes undesirable from the point of view of the isolated ethnic group, leading to an accumulation of frustration and hostility. The territory of the groups with the lower status can eventually turn into a 'no-go zone' for the elite or the state. For example, even policemen are hesitant over entering the *favelas* and do so only in massive operations. In autumn 2000, two Israeli soldiers who went astray in a Palestinian area (not in the Little Triangle) were lynched by the inhabitants in front of cameramen who later broadcasted the images to the world. In contrast, in the case of the affluent districts in Rio, Jakarta and Taiwan, the residential segregation is self-imposed. Such segregation provides protection to the groups concerned, at least for the time being. However, a social revolution could turn these districts into inescapable traps for their residents. Nevertheless, one positive effect of residential segregation is that places of worship established in these settlements function as centres of community life and may even have a beneficial effect on the health of the community.

The mingling of various groups, as observed in the luxurious Beirut apartment blocks, can reduce residential segregation and the tensions associated with it. Yet another process of eliminating residential segregation – not discussed further in this volume – is the situation when ethnic boundaries become blurred. This can happen either through the assimilation of ethnic minorities into mainstream culture (with the host culture also adopting, to some degree, the cultural traits of the minorities) or through mixed or intercultural marriages. Finally, where the ethnic/spatial division has been made on equal terms or on terms accepted by all the parties involved, such as in contemporary Beirut and in Tehran, the residential segregation can be bypassed by frequent encounters in public places. This aspect is discussed in the following section.

The appropriation or usage of public space

The distinction between 'private' and 'public' space becomes useful in analysis when adequately defined. Usually private space is equated with the built environment, where 'houses, stores, factories, churches, offices and schools usually have distinct physical characteristics that indicate their intended use'. This is contrasted with public streets and walks that take up (presumably in the USA) as much as 30 per cent of the developed land in urban areas (Jacobs 1985: 32, 67).

We believe, however, that instead of the function of a place, the criterion to be used in the demarcation should be the question of whether or not there is a party that unquestionably determines access to that space. From this position, shops, restaurants, schools and other buildings that provide public services in fact become private spaces, because usually one person literally holds the key to that particular space and determines how that space is to be used. Strangers may freely enter a restaurant, but it is only the proprietor or manager who determines the usage of space in it – that is, whether it is for dining exclusively, whether there is also a corner for drinks, whether there is a larger or smaller zone for non-smokers and so on. In contrast, 'public space' is freely accessible to the general public, with no restrictions – such as according to gender, age or ethnicity – outside those defined by law. Since public space cannot be claimed as private property, people do not even attempt to build anything permanent on it. If they did, the structure would probably be removed by others. Therefore, public space is almost always open space – streets, squares, sidewalks, parks and other vacant lots – with a limited physical infrastructure of pavements, gutters, trees providing shade, and perhaps some benches.

Because of the free access and the absence of any constructions restricting activity, the use of these public spaces is temporary and can change – even several times a day. For example, an urban park functions as playground for children in the morning, becomes a speakers' corner in the afternoon, an outlet for foodstalls during the evening and a sleeping place for tramps at night. In a built-up area, behaviour is largely structured by the function of the building entered, but behaviour in public space is regulated by rules of conduct more often than not circumscribed by one's culture. Since such rules even differ between subcultures, let alone between different national cultures, they can lead to embarrassment or misunderstanding when different ethnic groups interact. This is one reason why public space in multicultural settings can become an arena of social conflict between its potential users. Social conflict is enhanced when public spaces are appropriated by specific groups. The invisible borders of this private turf may be defended against outsiders (Colombijn 1994: 303–4; Lefebvre 1986 [1974]: 69; Oosterman 1993: 21–2).

Public space becomes a central issue when the existence of an ethnic neighbourhood is questioned in terms of social interaction, instead of (as Darden and others have done) in terms of the ethnic (residential) composition of the population. Greg Guldin (1985) suggests that an ethnic neighbourhood exists when one particular ethnic group experiences not only its primary relationships (weddings, close friendship and so on) but also its secondary relationships (shopping at retail stores, casual interactions and so on) as ethnically segregated. These social relationships can clearly be observed in public places, or 'interactional venues' as Guldin calls them. We believe that these two types of social relationship should be extended to include those beyond intra-ethnic contacts, since public space is first and foremost an 'interactional venue' of inter-ethnic contacts. Since even in a totally segregated 'plural society', where the constituent populations have neither a social life nor cultural norms in common, different ethnic groups do have a common economic motive and thus interact in the market (Furnivall 1944: 446–64). With his usage of 'market', J.S. Furnivall suggests a visible gathering of people at a certain spot during certain hours – that is, in a public space.

Public space turns into a meeting ground for ethnic groups who reside in different streets or neigbourhoods, provided they leave aside their differences. Both in Tehran and Beirut, the boulevards are meeting places for children from different ethnic backgrounds. In Tehran, people from different ethnic groups rarely visit each other at their homes but meet in public spaces. This is particularly so during special occasions, such as an important football victory or the annual ritual feast when people rejoice by jumping over a fire and lighting firecrackers. In Beirut, the Corniche – alias the seaside promenade – is a popular public space. However, Genberg (Chapter 5) has reservations about the plans for a sixty-metre-wide extension of the Corniche. He fears that it will then be so wide as to prevent eye-contact being made between pedestrians who are used to – in keeping with Middle Eastern behaviour – observing other people while walking and being watched by passersby – hence, the Corniche would cease to function socially. There has been one occasion, a special event in 1998, when all the groups in Beirut jointly made use of this public space, even if it was only for one night. On that particular night, many popular Arab singers attracted more than 100,000 spectators. SOLIDERE, the company responsible for the urban development of Beirut, had encouraged the concert to take place at this central square in order to prove to the residents that this square was to be the 'natural' meeting ground for all of the residents, regardless of ethnic origin.

In contrast to and behind this promising picture of a public space in postwar Beirut, linger the dreadful dynamics of space from the time when internecine turf wars between the militias of Beirut had been widespread. During that period the prominent streets and squares had lost their former identity for the worse, becoming awesome landmarks and demarcating lines. Formerly insignificant, winding alleys gained importance as relatively safe routes. The distinction between private and public space became blurred when basements, rooftops and windows in private homes were used as shelters or bunkers during shootings in the streets, while the roads were 'domesticated' through family belongings spilling on to the street to construct barricades. Semi-public spaces, such as verandas and balconies, were shuttered to protect them against snipers and shell-fire (Khalaf 1993: 36–7).

Perhaps one of the most hopeful chapters in this book is the analysis of a street festival that took place on a square in Vienna (Chapter 12). During the two days of the festival, many groups occupied the usually vacant square for a range of festivities. Groups, which in ordinary life avoided each other, had the opportunity to mix. Dumreicher uses the term 'emotional co-ownership', which expresses excellently the notion that people can control public space, but cannot own it as one can own real property. One familiar atmosphere uniting various ethnic groups was evoked by a carpet, chairs, a couch and a respectable woman (in fact, nobody else than Heidi Dumreicher herself) serving tea to her guests. Since it was positioned right at the centre of the square, it seemed as if the giant square had been reduced to a living room, open to anybody who would like to visit. Once the carpet was removed, because it had to be used as the backdrop for a peformance elsewhere, the unifying spell was broken. Later, the couch was also taken away by one of the chance 'actors' of this display and utilized as a seat to observe people in the square. Afterwards, when it was flung into a fountain, causing damage in the process, an aggressive atmosphere had emerged. Apparently, good inter-ethnic relationships cannot be easily engineered to be enduring.

Public space is not always a meeting place. It can also become a battlefield. In Albuquerque (Chapter 13), Anglo-Americans and Hispanics contested the meaning conveyed by the city centre, respectively defining it in economic or religious-cultural terms. For this reason the Hispanics opposed the sale of liquor at an Anglo-American restaurant situated opposite a Catholic church. The annoyance about the restaurant was more of an excuse to start a political fight than a statement related to a real issue, since brothels and taverns near churches have been quite common in the past. In the end, the proprietor of the restaurant got his licence, but not without a struggle that took several years. The struggle between these groups then shifted to a religious parish procession, for which the streets had to be closed. This time the Anglo-American shop owners complained about financial losses due to the closing of the street and the Hispanics interpreted their objections as showing a lack of respect for the traditional culture of the Hispanics. Once again the issue revolved around the question of whether the city centre was an economic area or a milieu for the practice of cultural and religious traditions.

State intervention in the allocation of space

The previous classification of the macro-, meso- and micro-level of analysis (applicable to the city, the neighbourhood and the street respectively) must not blur the fact that there is also a supramacro-level of analysis that focuses on the state and the international community. The Taiwanese from Hsinchu Science-based Industrial Park who lead a transnational life, form the most spectacular example (Chapter 9) in the volume. Another excellent example comes from Philadelphia Chinatown (Chapter 8), with the mobilization of Chinese support from coast to coast of the USA when Philadelphia Chinatown was threatened by a project proposing a baseball stadium, for which it would have been necessary to demolish part of Chinatown.

The state has the potential to play a very prominent role in inter-ethnic relationships. Oren Yiftachel argues that the state can influence such relationships by urban planning. Conventionally, spatial planning is seen as a tool aiming at modernist goals, such as providing local amenities, economic efficiency and, in general, pressing for change to a 'better' society. But planning also connotes being regressive and controlling. Three dimensions of planning – territorial, procedural and socioeconomic – come to mind as means of control. Each of these three dimensions facilitates the domination of a key societal resource: respectively, space, power and wealth. Territorial planning, the dimension related to domination of urban space, may be possible in three ways: through segregation, surveillance and the redrawing of boundaries in order to contain one group within a certain area (Yiftachel 1995). In fact, at the root of planning as a means of control lies the popular 'landscape of ... conflicting desires and fears' (Sandercock 2000: 202). Thus urban planning can also be seen as the management, or containment, of whatever fears exist in the city.

Following Yiftachel's argument, Abu Rass (Chapter 4) distinguishes between four possible planning strategies in ethnically divided societies. In the first, territorial issues are depoliticized by treating people as individuals rather than as members of ethnic groups; this strategy tries to avoid political exclusion of any kind. In the second, the planners of the dominant group act in a partisan manner to protect their own interests and usually those of the dominant group as a whole. In the third, the resources are allocated on an equitable basis, with the aim of minimizing any political and economic disparities between the ethnic groups. In the last, urban planning can be carried out in such a way as to promote mutual empowerment and tolerable co-existence for the parties concerned. In practice, these strategies are not always easily discernible in separation and are usually blended or combined.

The first strategy – depoliticizing space by treating people as individuals – has been adopted in Beirut, a city that allows one to cherish hope for peace in Jerusalem and the Little Triangle. In an attempt to avoid offending any possible ethnic sensitivities, complete outsiders – the French company SOLIDERE – were entrusted with the reconstruction of Beirut. A decision to contract this mission to SOLIDERE could have been interpreted as an admission of weakness on the part of the government were it not for the fact that SOLIDERE can only work within the guidelines set by the Lebanese governments. Thus the state is still involved, though indirectly, in determining spatial-ethnic relationships in the city.

The second strategy – partisan planning – has been in vogue in Israel. The state, dominated by the Jewish population, has clearly pushed back the Arabs to a limited space. State policy has been to give priority to building new Jewish towns, to the planning of the Trans-Israel Highway and to concentrating the country's land in state hands. In a renowned lawsuit, which has come to be known as the 'Qazir Case', an Arab family went to court in protest against discriminatory responses from both the Israeli Land Authority (ILA) and the local government of a new Jewish settlement. The ILA and local government had refused to sell or lease land to this Arab family, since the land was reserved for Jews. Because Arabs have been forced to live in old towns where any development has been circumscribed, land prices are much higher than in the spacious new Jewish settlements. Eventually, the Israel High Court ruled in favour of the Arab family, stating that the decision to refuse land had indeed been discriminatory; since then, however, there have been no endeavours by the state to implement this decision. The outcome of such discriminatory state policies in Israel has been unequal access to land and a segregated and embittered society.

Jerusalem may also serve as an example of partisan planning. In 1948 the city was partitioned between Jordanian and Israeli forces. After Israel reunited the whole of the city during the Six Day War of 1967, urban planning – and the accompanying reconstruction of some districts while razing others – sought to propagate the message that the city cannot be partitioned again and will remain under Israeli control (Friedland and Hecht 1996: 43). Another example of partisan planning can be seen in the major Brazilian cities where elite groups, through their control over urban planning, contain the lower classes in *favelas*, some of which are actually physically constrained by the walls – put up for protection – of the neighbouring elite residential complexes.

Partisan urban planning was also characteristic of colonial cities. Though Harald Leisch focuses on the Chinese in his chapter, rules specifying which locations were reserved for certain groups were pertinent to all ethnic groups in Dutch colonial Indonesia. In the British colonial city of Singapore too, bylaws reinforced ethnic-spatial relations. In this former colonial town, the oldest masterplan marking out ethnic neighbourhoods dates back to 1823, only four years after the town was established. However, since the 1960s Singapore has passed on to an advanced and depoliticized planning strategy. Thus the state housing policy of building public housing for the majority of the population and distributing apartments on a 'first come, first served' basis has resulted in a desegregated residential pattern there (Chiew 1993: 288–92; Colombijn 1994: 50–63; Dale 1999: 14; Wertheim 1956).

The third strategy – allocating resources on an equitable basis – seems to have been adopted by Canada, a nation-state officially encouraging the development of its multicultural population. However, the chapters by Darden and Shah present a different picture of urban planning in Canada. The national multicultural policy does not seem to be manifested in local and urban spatial planning. In other words, this national policy has not prevented urban ethnic segregation.

Similarly, in the USA the principles of official policy are at odds with policy in practice, according to a critical article by Tanya Y. Price. In Washington, D.C., the constitutional equality contrasts with the ethnic-spatial inequality. The Federal enclaves of offices, courts and museums constitute a predominantly 'white' public space in the middle of a predominantly 'black' city. The ethnic tension in Washington is exacerbated by tension between the Federal and District (local) governments. The urban planning department delivers poor services to neighbourhoods with a high proportion of non-white residents. For example, a metro line that links two 'white' suburbs with each other and the central city has been operating for over a decade, but another line that should be connecting a lower- income and predominantly 'black' neighbourhood is still under construction (Price 1998). All in all, despite the rhetoric of equality, in many respects Washington resembles the hierarchic Brazilian cities, including the similarity of direct transportation between the places claimed by the elite. It would be fair to think that in spite of these negative depictions, there are surely also examples of urban planning based on equality that have had positive results – like the one in Singapore mentioned earlier.

The fourth strategy – mutual empowerment and tolerable co-existence – was pursued in Vienna, though not by the state. Instead, it was pursued by a local NGO that organized the Waterworld Festival (Chapter 12). Stretching the point a bit, the fourth strategy can also be recognized in Japan. For a very long time, Korean immigrants were considered to be outsiders who either had to assimilate or face exclusion. In the 1990s, after an extensive debate by politicians, scholars and activists, a new concept of 'living together' and accepting cultural difference was formulated. Though no concrete policy can be deducted from the concept of 'living together' (with difference), it broadly sanctions the new attitude of local governments to the Korean presence in Japan. In Osaka, this new concept was instrumental in creating an atmosphere in which a local merchants' association could launch a project for 'Korea Town', through which Korean culture was tolerated and stimulated (Chapter 11).

The strategy of empowerment and co-existence is also emerging, though slowly, in what were once the most notorious examples of ethnically divided urban settlements: South African cities. The Group Areas Act of 1950, preceded by legislation dating back to 1913, designated special areas for the four racial categories – blacks, whites, coloureds and Indians – and envisaged moving the black townships that were deemed to be too close to white areas to different sites. The implementation of this Act involved enormous financial, social and psychological costs and resulted in an almost totally segregated society that reflected a 'racial' social inequality of immense proportions. Yet, as far back as the 1980s, the blacks had begun to squat the vacant properties in the white neighbourhoods of Johannesburg. By 1991 the Group Areas Act had been repealed. The eviction of black squatters came to an end and many of these settlements were given legal status. In brief, the new policy gave birth to some hitherto unknown developments: a steady rural-to-urban migration of blacks; a black presence in the previously white city centres, especially those of Johannesburg and Durban; the desegregation of the suburbs, mostly due to legalized black squatter settlements; and the black occupation of pockets of vacant land formerly used as buffer zones between the 'racial' groups. Another outcome, reminiscent of what happened in Southeast Asia when the European colonial powers withdrew, was that a new spatial division based on class became discernible. At first, the technocratic town planners in South Africa had willingly implemented the Group Areas Act, but even before 1991 they had became aware of the political nature of their work. To what extent the new planning strategy will result in better social relations between the ethnic groups in South Africa remains to be seen. It is worth noting that at Hillbrow in Johannesburg, an almost desegregated neighbourhood, the former pattern of racial contact has survived desegregation intact, with relationships remaining within one's own group in spite of the new inter-ethnic residential proximity. Indeed, the recent influx of black West Africans to Hillbrow provoked a xenophobic response, not only from the white inhabitants but also from the South African blacks in the area (Khan and Maharaj 1998; Morris 1999; Saff 1998: 45–80).

The fluidity of ethnic identity

Contrary to popular belief, ethnic identities are not fixed. In fact, in the short term, they are situational and change with the interlocutor that a person is facing. In the long term, identities can gradually develop when groups are pitted against each other. In the latter case, ethnicity may be influenced by spatial changes. When immigrants come to a city (or a country) and decide to live in an 'ethnic urban village', their identities change. The place will force an ethnic identity on the people concerned. The new ethnic identity will provide some protection and natural allies, but at the same time it will block access to certain opportunities. The new arrivals in Philadelphia Chinatown and the Vietnamese immigrants in Toronto serve as examples here.

Urban space can also have an impact on ethnic identity in the short term. In the previous section several examples were presented of special occasions when identities could be merged in public places. The Fire Feast is an Iranian rite that offers the opportunity for different groups to share a common experience in a public space. During one evening of the year, the four groups that constitute the district of Ekhtiarieh in Tehran come together on the boulevard and join a communal celebration around the bonfires. But the procession of the month of Moharram, which as a Muslim celebration has the potential to unite all groups in the neighbourhood, is monopolized by one particular group to the exclusion of the other Muslim residents of Ekhtiarieh. This group, originating from the Kashan region outside Tehran, temporarily takes over the public space. This confirms Gerd Baumann's conclusion that 'people manage to shape dialogical identities while at the same time reify monological ones' (quoted by Shahshahani in Chapter 10). Also, Genberg provides us with the example of an open-air concert in the central square of Beirut uniting all groups, even if only for one night.

An interesting example of the fluidity of ethnic identity is provided by Chang. Taiwanese workers in information technology have a transnational lifestyle, moving back and forth between Hsinchu Science-based Industrial Park in Taiwan and Silicon Valley in the USA. Their innate identity remains, however, remarkably stable. It is only in Taiwan that this group's outer identity is also marked by class – as the affluent 'Parkers' – whereas in California it is determined by ethnicity. The *Ranch 99* chainstores in Silicon Valley play a central role in their ethnic identity, enabling them to maintain that identity in the USA. It is interesting to note that non-Taiwanese Americans do not recognize the *Ranch 99* stores as anything special, but the Taiwanese immediately identify them as being typically their own place.

Ethnic design in the cityscape

The final thread that runs through many chapters and is worth noting, concerns the role space or its usage plays in creating a medium for the display of ethnic symbols or emotions. As was mentioned in the introduction, space can be instrumental in conveying symbolic meaning, as well as being an object of production or consumption (Lefebvre 1986 [1974]). Studies undertaken in the Copperbelt towns of Zambia in the 1940s, 1950s and 1960s, showed that urban migrants used various strategies to communicate cultural difference. In some situations, ethnicity was emphasized: it was shown off in public rituals. In such a case ethnic identity was, using Goffman's term, 'overcommunicated'. In other situations, ethnicity was played down or 'undercommunicated' (Eriksen 1993: 20–2; Hannerz 1980).

The chapters in this volume suggest that there is a remarkable inverse relationship between the strength of inter-ethnic antagonism and the visibility of ethnic symbols. That is, in places where inter-ethnic relations are tense and emotional, ethnic symbols are demoted and confined to keeping a low profile – they are 'undercommunicated'. In contrast, where inter-ethnic relations are good or where ethnic groups have become integrated with each other, ethnic colours are publicly flown – 'overcommunicated'.

In this context, ethnicity is regarded as a positive asset and openly exploited as a means to promote shopping streets in Osaka, Philadelphia and Albuquerque. The similarity of ethnic flavouring used for the street lamps, entrance gates and pavements of Korea Town in Osaka and Chinatown in Philadelphia is striking. However, in such cases ethnicity is merely expressed as decorative objects, which are 'non-authentic' in some cases. Of course, 'authenticity' can be, in a way, a misleading concept. It suggests that an object exists in an original and unspoilt form. Such a static view of cultural artifacts is untenable. The term 'non-authentic' is used here to refer to a particular form that has been adjusted in relation to a new object that is alien to the culture from which that form originated: for instance, a pagoda roof on a telephone booth. Yet the essential point under discussion here is not the authenticity of the ethnic decorations, but the fact that overt cultural expression by ethnic minorities is only allowed when it is merely a decoration in outward appearances.

Elite neighbourhoods are, in contrast, 'de-ethnicized' and display an American way of life. This way of life extends to architectural styles. In Jakarta, the real-estate developers can market houses by giving them striking Western names because, as one manager remarked, the customers 'aren't too keen on ethnic [Indonesian] architecture … . They want "Mediterranean style", "Los Angeles style" and "Beverly Hills style"' (Kusno 2000: 113).[4] This de-ethnicized style can be perceived in such diverse places as the Hsinchu Science-based Industrial Park in Taiwan; the new towns surrounding Jakarta; the suburbs to which the newly affluent residents of American Chinatowns flee; and the condominiums in Brazil. A closer look at such settlements reveals that the resemblance is in outward appearance only. The lifestyles in these elite neighbourhoods can differ considerably. For instance, the Chinese in the Hsinchu Science-based Industrial Park lead a transnational life, but when in the USA they prefer to frequent a suburban Chinese chain of shops. Nevertheless, while the other Chinese in America satisfy their ethnic taste by making sentimental journeys to Chinatowns, such trips are despised by the 'Parkers' from Taiwan. Another example of difference in lifestyle among the inhabitants of these de-ethnicized elite neighbourhoods originates from affiliation to different religious denominations, as seen among the various Chinese elite groups and the Brazilian elite.

The reverse relation between ethnic antagonism and the visibility of ethnic symbols seems only to be true in cases where conflict over the ethnic-spatial situation has been settled. In those cases where public space is still being contested and the outcome of the contest is as yet undetermined, different groups will intentionally fly their colours at hot spots. This is exemplified by the vexatious Orange marches in Northern Ireland, where Loyalist Protestants carry provocative orange banners and wear British-style uniforms. The marches commemorate the Battle of the Boyne, which took place on 12 July 1690, and continue throughout the summer. Some of these marches, but not all of them, deliberately take place on streets known to be inhabited by Catholics; the marchers also take care to drum loudly in front of Catholic churches (Ruane and Todd 1996: 108–10).

Similarly, ethnic symbols can be openly and extensively displayed by ethnic majorities in spaces allocated to this purpose. One example is the Indonesian city of Padang, which has a population consisting of indigenous Minangkabau – who form the majority – and a minority population of Chinese along with other ethnic groups. The Minangkabau-dominated municipal administration ordered that all buildings of three floors or more be adorned with the quasi-traditional Minangkabau saddle roof, with pointed horns at the end of the roof-beams. Similarly, saddle-roofed gates were placed at all entrances to the city: on the major roads, in the harbour and at the airport. The result of this local bylaw for buildings was a striking cityscape of pointed saddle roofs. One reason for the mushrooming of these saddle roofs was inter-ethnic tension that led the administration to attempt to claim the city as Minangkabau territory, at least at a symbolic level. The Chinese presence in the city, most visible in the Chinese graveyard on a hill overlooking the city centre, also had to be removed: the Minangkabau administration ordered that it be moved to a less conspicuous place. After a conflict that lasted for years, the graveyard was finally demolished. So, in the case of Padang, ethnicity has had a major impact on the form of urban space, but not so much on its residential segregation (Colombijn 1994: 335–7).

A recapitulation of the major findings of this volume would be as follows: Residential segregation has a negative impact on the chances of equality in development, when that segregation is imposed by a hegemonic group on a weaker one. An example from North American cities is the segregation of African-Americans based on the ideology of white supremacy. A corollary of the white supremacy ideology is that, when ethnicity is taken as being ascribed and circumscribed by racial (that is, hereditary) difference, it freezes ethnic boundaries. Two countervailing inferences can be drawn from residential segregation: One is the beneficial health effect of living close to a prayer house of one's own denomination, while the other is the mutual protection, if not support, provided by living among members of the same ethnic group. Fear of violence from the lower classes, accompanied by ethnic difference, can lead to the residential segregation of the elite in gated communities where leaders of different ethnic groups – sharing the same negative attitude towards the lower strata – can come into contact with each other on common grounds. Public spaces are important meeting grounds for people of different ethnic backgrounds from all classes. During these meetings, ethnic identity is fluid and can be temporarily exchanged for an encompassing identity. Public space, however, may also be contested. The state can intervene in the allocation of space with varying degrees of neutrality – or by playing a partisan role. Ethnic designs are used in commercial areas, where ethnic tension is low. In places where tensions are high and in elite neighbourhoods, ethnic symbols are downgraded – except where they are meant to be provocative.

As a final word, we would like to remind the reader that all of these findings – as we have already stressed in the introduction and at the beginning of this chapter – are empirical generalizations based on a limited number of cases. Further research on the relationship between space and ethnicity, with further case material coming from around the world, is urgently needed to arrive at healthy conclusions.

Notes

1 Ethnic segregation is easier to measure than the quality of inter-ethnic relations. Joe Darden (Chapter 2) gives a simple formula for segregation, which can be applied to any culture if the necessary data are avalaible. An easy measurement of the quality of inter-ethnic relations still does not exist. This is mainly because of it being a composite variable, with the meaning of each constituent part differing from culture to culture. Blows exchanged between furious Anglo-Americans and Hispanics in Albuquerque, for instance, may be a graver insult to the Hispanics – known for having a culture of honour that would necessitate reprisals – than to the Anglo-Americans.
2 In keeping with the statement of the International Union of Anthropological and Ethnological Sciences on 'race' (http://ruls01.fsw.leidenuniv.nl/~nas/08-race.htm; downloaded 7 November 2001), we would like to point out that all human beings belong to one species; that races in the sense of ethnically homogeneous populations do not exist; that it has not been possible to discern discontinuity in any pattern of genetic variety and that there is no hereditary justification for considering one population superior to another.
3 At the root of this ethnic economic specialization lies the notion that mutual trust, which is the basis of cooperation, easily flourishes between people from the same ethnic background. The positive effect of trust overcomes the fact that the concentration of one ethnic business in one urban quarter is against the economic logic that each business seek its own clientele by keeping at a maximal distance from similar enterprises.
4 In order to counter such globalizing trends, in 1995 the Indonesian government ordered the replacement of English names with Indonesian ones. This ban on English was, in fact, associated with unspoken suspicions about the national loyalty of Chinese immigrants (Kusno 2000: 156–8).

References

Arbide, Dardo, Marra, Guillermo and Tavormina, Sebastian (2000) *Transformaciones en los espacios publicos de comida: Mapas del imaginario gastronómico* (Transformations of Public Space for Eating: Maps of the gastronomical imagination), Buenos Aires: Belgrano.

Chiew Seen Kong (1993) 'Towards a theory of transition from ethnicity to class: ethnic relations in Singapore', in Lee Boon Hiok and K.S. Susan Oorjitham (eds), *Malaysia* and Singapore: Experiences in industrialization and urban development, Kuala Lumpur: University of Malaya.

Colombijn, Freek (1994) *Patches of Padang: The history of an Indonesian town in the twentieth century and the use of urban space*, Leiden: Research School CNWS.

Dale, Johan (1999) *Urban Planning in Singapore: The transformation of a city*, Oxford: Oxford University Press.

Darden, Joe T. (2001) 'Race relations in the city', in Ronan Paddison (ed.), *Handbook of Urban Studies*, London, Thousand Oaks and New Delhi: Sage.

Eriksen, Thomas Hylland (1993) *Ethnicity and Nationalism: Anthropological perspectives*, London and East Haven: Pluto.

Fong, Eric, Luk, Chiu and Ooka, Emi (2000) 'Spatial distribution of suburban ethnic businesses', paper presented at the IUAES 2000 Inter-Congress, Beijing, 24–28 August.

Friedland, Roger and Hecht, Richard (1996) *To Rule Jerusalem*, Cambridge: Cambridge University Press.

Furnivall, John Sydenham (1944) *Netherlands India: A study of plural economy*, Cambridge: Cambridge University Press.

Guldin, Greg (1985) 'Measuring urban ethnicity', in Aidan Southall, Peter J.M. Nas and Ghaus Ansari (eds), *City and Society: Studies in urban ethnicity, life-style and class*, Leiden: Institute of Cultural and Social Studies, Leiden University.

Hannerz, Ulf (1980) *Exploring the City: Inquiries toward an urban anthropology*, New York: Columbia University Press.

Jacobs, Allan B. (1985) *Looking at Cities*, Cambridge and London: Harvard University Press.

Jenkins, Richard (1996) *Social Identity*, London and New York: Routledge.

—— (1997) *Rethinking Ethnicity: Arguments and explorations*, London: Sage.

Jezernik, Bozidar (1999) 'The old bridge of Mostar', in Bozidar Jezernik (ed.), *Urban Symbolism and Rituals*, Ljubljana: Univerza v Ljubljani.

Khalaf, Samir (1993) 'Urban design and the recovery of Beirut', in Samir Khalaf and Philip S. Khoury (eds), *Recovering Beirut: Urban design and post-war reconstruction*, Leiden: Brill.

Khan, Sultan and Maharaj, Brij (1998) 'Restructuring the apartheid city', *Urban Forum* 9: 197–223.

Kusno, Abidin (2000) *Behind the Postcolonial: Architecture, urban space and political cultures in Indonesia*, London and New York: Routledge.

Lefebvre, Henri (1986, orig. 1974) *La production de l'espace* (The Production of Space), 3rd edn, Paris: Éditions anthropos.

Lucchesini, Alessandro (1993) *Cinesi a Firenze: Storia e biodemografia di una colonia di immigrati* (Chinese in Florence: History and demography of a colony of immigrants), Florence: Angelo Pontecorboli.

McCarthy, John (1999) 'Chicago: a case study of social exclusion and city regeneration', *Cities* 16: 323–31.

Morris, Alan (1999) 'Race relations and racism in a racially diverse inner city neighbourhood: a case study of Hillbrow, Johannesburg', *Journal of Southern African Studies* 25: 667–94.

Oosterman, Jan (1993) *Parade der passanten: De stad, het vertier en de terrassen* (Parade of Pedestrians: The city, the fun and the sidewalk cafes), Utrecht: Van Arkel.

Paddison, Ronan (2001) 'Communities in the city', in Ronan Paddison (ed.), *Handbook of Urban Studies*, London, Thousand Oaks and New Delhi: Sage.

Phillips, E. Barbara (1996) *City Lights: Urban–suburban life in the global society*, 2nd edn (1st edn 1981, E. Barbara Phillips and Richard T. LeGates), New York and Oxford: Oxford University Press.

Price, Tanya Y. (1998) 'White public spaces in black places: the social reconstruction of whiteness in Washington, D.C.', *Urban Anthropology* 27: 301–44.

Quispel, Chris (1996) ' "Amerikaanse toestanden" in Nederland?' ("American troubles" in the Netherlands?), *Tijdschrift voor Sociale Geschiedenis* 22: 327–47.

Ruane, Joseph and Todd, Jennifer (1996) *The Dynamics of Conflict in Northern Ireland: Power, conflict and emancipation*, Cambridge: Cambridge University Press.

Saff, Grant R. (1998) *Changing Cape Town: Urban dynamics, policy and planning during the political transition in South Africa*, Lanham: University Press of America.

Sandercock, Leonie (2000) 'Negotiating fear and desire: the future of planning in multicultural societies', *Urban Forum* 11: 201–10.

Sennett, Richard (1973, orig. 1970) *The Uses of Disorder: Personal identity and city life*, Harmondsworth: Penguin.

Wertheim, Willem Frederik (1956) 'Urban development', in Willem Frederik Wertheim, *Indonesian Society in Transition: A study of social change*, The Hague and Bandung: Van Hoeve.

Yiftachel, Oren (1995) 'The dark side of modernism: planning as control of an ethnic minority', in Sophie Watson and Katherine Gibson (eds), *Postmodern Cities and Spaces*, Oxford and Cambridge: Blackwell.

Index

A'ra 73—4
A'ra'ra 73–4
Abu Rass, Thabet 15, 237
Acre 65
Afghanistan 166–7
Afghans 18, 160–1, 165–6, 170
Africa 228
African-Americans *see* blacks
Africans 27, 110, 116; West 239
age 51
Agenda 21 193
Ahl-e Haq sect 170
Albuquerque 19–20, 209–23, 236, 241
Aleppo 84
Amal 84
Amazonia 109–10
Ambon City (Kota Ambon) 11–13
American movies 117
American style architecture 16, 106, 144–7, 241
American suburban standard 17
American way of life 16, 106, 149, 241
Americans 46, 127, 130, 139, 150, 152, 220
Amerindians 19–20, 110, 209–11, 213–18, 220, 222–3; Navajo 213–14, 217, 222; Pueblo 211, 213–14, 217, 222; Zuni 214
Anglo-Americans 19–20, 209–13, 215–20, 222, 236
anonymity 206
apartheid 119; *see also* South Africa
Appadurai, Arjun 9
Arab towns 61–2, 67–78
Arab–Israel War of 1948–1949 64, 77, 237
Arabs 15–16, 31, 61, 63–79, 91, 116, 227, 232, 235, 237
Arendt, Hannah 92, 94 n. 15

Aristotle 93, 95 n. 18
Arizona 215
Armenians 84
arms 122
artists 202, 211
Asia 228
Asians 27, 52, 129, 135, 142, 153, 202, 213, 228
assimilation 5, 107, 128, 183, 187–8, 227, 233; forced 101; spatial 28
Augé, Marc 117, 209
Australia 101, 149
Austria 19, 199; *see also* Vienna
Austrian 206

Baha'i 52
Bali 102
Baltimore 136
Bamboo Village 17, 143, 145–6, 148–50
Bandar Abas 160, 166–7
Bangladesh 3
banning of public display of Chinese culture 16, 101, 103
Baqa al-Gharbiyye 70, 73, 77
Barak, Ehud 61, 78
barriers, cultural 200
Barth, Fredrik 10
Battle of the Boyne 242
Baumann, Gerd 161–2, 172, 240
beach 120
Beirut 1, 15, 18, 81–93, 227, 232–5, 237, 240; Corniche at 90–2, 235
Belém 110, 125 n. 18
Belfast 1
Belo Horizonte 110
Ben-Gurion, David 65
Benjamin, Walter 117
Berlin 1
Berri 84

bi-polar space 117
bilingual people 104, 195; school for 138, 145; street signs 134, 136–7, 154, 186
blacks 17, 28–31, 51, 81, 111–13, 116–17, 119, 121–3, 130, 213, 228, 238–9, 242
blocking expansion of Arab settlements 67
Boal, Augusto 201
Bolivia 110
Bosnia-Herzegovina 227
Boston 1, 137, 145; West End 1
boulevard 18, 145, 166–7, 171, 234, 240
boundaries 70, 81–2, 86, 88, 93, 138, 160, 165, 170–2, 177, 201–2
bourgeoisie 107
brain-drain 148
Brasília 109
Brazil 16–17, 109–23, 229–32, 238, 241; regions in 109–10, 118
Brazilian 204
British 17, 29–30, 128, 130, 170, 238
Buddhist temple 53, 132
Buddhists 14, 36, 52–3
Buenos Aires 230
Bumi Serpong Damai 105–6
Burgess, Ernest 5, 8, 15, 229
Burmese 35
bus station 120
business centre 130; *see also* central business district

California 17, 48, 50, 142, 151, 154; *see also* Silicon Valley
Cambodians 35, 132
Canada 27–31, 33, 43 n. 1, 46, 52, 101, 110, 122, 149, 229; *see also* Toronto; Vancouver
Canadian citizenship 35
Canadians 30, 33, 36, 42
Cantonese 134
Cape Town 1
Caribbean people 27
Castells, Manuel 2
Caucasian *see* whites
central business district 5, 86, 89, 126, 138
Chang, Morris 155
Chang, Shenglin 17, 230–1, 240
Cheju 178, 180
Chicago 1, 5, 17
Chicago School 1, 5–8, 227
children 12, 18, 109, 113, 116, 121–3, 123 n. 3, 125 n. 20, 134, 137–8, 148–50, 154, 165, 167–72, 194–5, 200–1, 232, 234
Chin-shan-mian 145–6, 149, 157 n. 2
China *see* People's Republic of China; Taiwan
Chinatown 17–18, 103–4, 126–39, 152–4, 184–5, 187–8, 227, 229–31, 236, 240–1; in Indonesia (*pacinan*) 101
Chinese 13, 16, 17, 30–1, 99–107, 126–39, 150, 153–4, 187, 194, 227, 229–31, 238, 242; architecture 137; banning of public display of Chinese culture 16, 101, 103; food 103–4, 135, 142, 153; goods 104, 154; identity 227; political exclusion in Indonesia 99; temple 147, 231; *see also* Taiwanese
Christians 12–13, 15, 52, 56, 65, 83–7, 101, 227; *see also* Greek Orthodox; Maronites
church 12–13, 48, 51, 84, 120, 127–9, 131–5, 138–9, 197–8, 203, 210, 212, 219, 220–1, 229–30, 236
Churchill, Winston 3
Città degli villaggi movement 205
civic inequality 61
civil war in Lebanon 82, 85, 87, 91
class *see* socioeconomic class
Cleveland 136
Coen, Jan Pieterszoon 102
Cohen, Amnon 64
collective identity 64, 152
Colombia 110
colonial times 16, 101–2, 104, 107, 116, 187–8, 210, 219, 238–9
colonialism 4
coloured people *see* visible minorities
comfort zone 126
commodification of culture 188, 214, 217
communist states 2
community organizations 48, 57, 129, 131, 133, 136
competition for urban space 1, 5, 6, 8, 209, 217, 220, 220
condominium 16, 92, 115–8, 121–2, 238, 241
Copperbelt towns 9, 240
Corniche 90–2, 235
cosmopolitan urbanites 82, 169
Cressey, Paul G. 6
crime 29
criminal networks 11

248 *Index*

Croats 227
cross-cultural comparison 20
cultural difference 181, 183–4, 186, 188–9
cultural stereotype 223
Cupertino 151, 154
Curitiba 110
Czechs 194

Darden, Joe 14, 228–9, 234, 238, 243 n. 1
Davis, Mike 232
decolonization 2
decoration of the street 200; *see also* ethnically marked decoration
Delaware 138
Denver 136
desire 201
development 74; planning 65–7; projects 128–9, 132, 139
discrimination 46, 74, 76–8, 82, 127–8, 133, 136, 211, 217–19
domestic workers 116, 166
Druzes 65, 83
Dumreicher, Heidi 19, 235
Durban 239
Dürr, Eveline 19
Dutch East India Company (VOC) 102
Dutchmen 102, 104, 238

education 139, 154, 169–70
Ekhtiarieh 162–5, 167–70, 240
elderly 113, 121, 123, 130, 132, 134, 137, 198
elections 12
Elias, Norbert 113
elite 9, 12, 16, 39, 41, 99, 104, 112, 114, 117–19, 122–3, 231–2, 238, 241–2
elite residential complexes 16
emancipatory urban development 196–7
emigration 101
emotional co-ownership 197, 200–1, 235
environmental impact 147; *see also* pollution
Esfahan 160, 162, 166–7
ethnic architecture *see* ethnically marked decoration
ethnic boundaries 10, 13; fluidity of 13; fixation of 13; rearticulation of 222
ethnic business *see* ethnic economic specialization
ethnic categorization 17, 27, 116, 229–30; in Lebanon 81; *see also* religion, denomination

ethnic cleansing 9, 227; *see also* religious homogenization of space
ethnic community 126
ethnic composition 61
ethnic conflict 63, 154, 207, 214, 227–8, 238, 241
ethnic contact 18, 20, 196
ethnic consciousness 17, 126, 135, 137
ethnic diversity 4, 8, 10, 154, 179, 192–3, 199, 207, 211, 226, 233
ethnic economic specialization 17, 35, 230, 43 n. 3, 103
ethnic enclave 10, 13, 25, 46, 81, 88, 93, 116, 126–7, 129, 152, 230, 232, 234, 238, 240; in a suburb 230; *see also* moving out; multi-ethnic neighbourhood
ethnic groups 177, 194, 202, 227, 234–5; in Iran 160–1; spatial distribution of 38 (*see also* residential segregation)
ethnic homogenous space *see* residential segregation
ethnic identity 3, 6, 9, 10, 127–8, 130, 135, 137, 161, 188, 209, 239, 241–2; double 195; fluidity of 239–40
ethnic integration 15, 207; *see also* residential desegregation
ethnic intermarriage *see* intermarriage
ethnic melting pot 1, 4, 228
ethnic minorities 46, 63–5, 82, 99–100, 136, 209, 241; *see also* religious minority
ethnic personal names 101, 189 n. 3
ethnic neighbourhood *see* ethnic enclave
ethnic solidarity 130, 137
ethnic tension *see* ethnic conflict
ethnic urban village *see* ethnic enclave
ethnically marked decoration of street 19, 126, 130, 134, 137, 181–2, 188, 226, 240–1, 243
ethnically marked food 135, 181, 190 n. 9; *see also* Chinese food, Korean food
ethnically marked goods 18–19, 104, 131, 181, 214–16, 222; *see also* Chinese, goods; Korean goods
ethnicity 3–5, 9–12, 29, 51, 63–4, 82, 171, 226, 228–9, 231–2, 234, 240–1, 243; constructivist view of 9–10; instrumentalist view of 9; primordialist view of 9–10, 15; and social class 39–42, 74, 109, 111, 123, 172, 226, 230–3, 242; symbolic signs of 10, 16–17, 240, 242–3

ethnocide 3
ethnoscape 2
Europe 13, 161, 166, 193, 219; Eastern 122; Western 226
European cities 206–7
Europeans 52, 101–3, 105, 107, 220, 223; Northern 30
expressway 120, 130–1, 133, 135, 139

Far East 169
Fars 160, 162, 170
favela 16, 109, 111–12, 118–19, 124 n. 4, 232–3, 238
fear 12, 16, 29, 86, 100, 105, 107, 109, 112–14, 118–19, 121–3, 206, 231–2, 236, 242; culture of 109, 114–15, 118, 121; space of 86–7
federal government *see* government, central
feng shui 106, 154
festivals 207
feudal system 110
Filipinos 6, 31
First World 122
Florence 230
Florianópolis 109–10, 121
football 76, 166, 171, 235
forced relocation 146
forced urbanization 102
forces of production 6
Fortaleza 110
France 122, 156
Fremont 142, 154
French 90, 170
frontier settlements 68, 70, 72
fundamentalism 203
Furnivall, J.S. 234

Gabeira, Fernando 112
Galilee 65–6
Gans, Herbert 1, 8
gated communities 82, 103–7, 113, 115–17, 122, 145, 148, 231–2, 242
Gay Parade 207
Gaza Strip 13, 61, 65–6
Geertz, Clifford 9
Gemmayze 86–7
Genberg, Daniel 15, 235, 240
gender 121, 123, 170, 196; different mortality rate 48, 50–1; and violence 121–2
gendered space 8
generation conflict 204

genocide 3
gentrification of neighbourhoods 2, 128, 231
Georgia 48
Germans 17, 29, 110, 128, 165, 170, 195
Germany 122, 156, 166
Ghaffari family 162–5
ghetto 2, 5, 8, 17, 81–2
global arms market 122
global comparison 4
global network society 2, 13
globalization 3, 127, 198
Glodok 16, 102–4, 231
Goffman, Ervin 241
government: central 28, 61, 63, 65, 67–8, 70, 72, 74–6, 79, 101, 104, 107, 210, 237; local 19, 56, 67–8, 70, 72–7, 79, 116, 118, 133, 135–6, 138–9, 154, 179, 184, 186, 194, 197–9, 201–2, 204–5, 207, 215, 220–1, 237, 242
government land policy 63, 65, 75, 78
grassroots movement 7, 9
Great Depression 128
Greek Orthodox 84, 87
Greeks 29, 194
Grossi, Miriam Pillar 16
Guan, Jian 17, 229–30
Guangzhou 103
guarded buildings 16, 117; *see also* gated communities
guest workers 3
Gujarati 56
Guldin, Gregory 234
Gutmann, Rudolf 194, 199

Hadera 68
Haifa 65, 67
Haiti 113
Hakanese 145–7
handicrafts *see* ethnically marked goods
Hariri, Rafiq 84, 94 n. 8
healing of a divided city 89
health 14, 46–7, 49, 51–2, 57, 118, 211, 229, 233, 242
hegemonic ethnic group 33, 42, 70, 233, 242; *see also* whites, ideology of supremacy
Herat 162
Hester, Jeffry 18
high-tech jobs 142–3, 145, 147–8, 150, 152–6, 230
Hillbrow 239
Hindu temple 18, 54, 56, 229

250   *Index*

Hindus 52–5, 101
Hippies 215, 218
Hispanics 19–20, 28, 209, 211–5, 217–23, 236
Historic District 212–5, 217–20, 222
Hizballah 84
Ho, Arturo 130
Hollanda, Chico Buarque de 123
home 41, 113, 143, 145–6 152–4, 156, 157 n. 11, 166, 205, 215, 220, 223; decoration 151; ownership 33–4, 128, 228
homelessness 124 n. 8
Hong Kong 3, 154; people coming from 52
Honshu 177
hotels 120
housewives 142–3, 148–51, 157 n. 6
housing 7, 9, 75, 105–6, 129, 131–5, 165, 178; public 238
Hsinchu City 147–9, 151, 155, 158 n. 14
Hsinchu Science Park (Hsinchu Science-based Industrial Park, HSIP) 17–18, 142–52, 155–6, 231–2, 236, 240–1
human ecology 7, 30; *see also* invasion; succession

immigrant reception area 52
immigrant settlement 27
immigrants 3, 11, 17–18, 30, 35, 52, 54, 61, 64, 67, 72, 75, 77, 110, 127–9, 130–2, 135–7, 143, 152, 154, 156, 167, 170, 183–4, 192–4, 201, 206, 219, 227, 230, 238, 240; order of arrival of 227–8
immigration 36, 52, 100–2, 180, 211; United States Immigration Act of 1965 129
India 156
Indian Ocean 102
Indians 16, 52, 56, 154, 203, 239 239; American *see* Amerindians
individuality 200
Indonesia 7, 11–12, 16, 99–102, 232, 238, 242; *see also* Jakarta
industrial park 73, 145; *see also* Hsinchu Science Park
industrialization 110, 154, 156
information age 145; city 167; technology 155; transnational 156
intermarriage 99, 110, 165, 170, 190 n. 4, 227, 233
intifada 61

invasion (human ecology) 1, 126, 180, 203, 218, 222, 227
Ipanema 118
Iran 18, 160, 162, 164, 166, 168–9, 171, 240; provinces of 160; *see also* Tehran
Iran–Iraq War 163, 168
Iranian Revolution 163, 169
Irish 17, 128
Islam 11, 18; *see also* Muslims
Israel 15, 50–1, 64, 66–7, 69–70, 75–9, 82, 114, 122, 233, 237; Agricultural Settlement Law of 66; *see also* Little Triangle
Israeli Association of Citizens' Rights 76
Israeli Land Administration (ILA) 66, 76, 237
Italians 1, 29, 110

Jacareí 111–12, 124 n. 6
Jacobs, Jane 2, 8, 90, 94 n. 12
Jaffa 65
Jains 52
Jakarta 12–13, 16–17, 99, 102–7, 231–2, 233, 241; colonial (Batavia) 102, 105; Greater Jakarta (Jabotabek) 105
Jamaicans 29
Japan 3, 18–19, 122, 156, 177–9, 181–9, 238–9
Japanese 17, 19, 31, 110, 116, 177–81, 183–9, 229–30
Java 101
Jenin 68
Jerusalem 1, 13, 15, 67, 237; East 64; partition of 237
Jewish Agency 76
Jewish towns 62, 66–8, 71–6
Jews 5, 15, 30, 50–2, 63, 65–8, 70–9, 116, 227, 237
jihad 13
Johannesburg 1, 239
Jordan 69, 237
Judaization 63, 66, 77
Judeo-Christian faith 52, 54
Jumblatt 84

Kafar Qasem 68, 70, 77
Kashan 160, 162–4, 167, 169, 240
Kataaib party 84, 86
Khalaf, Samir 15
kibbutzim 50
kinship 7
Kobe 184, 187
Komanchi 160

Korea 178, 186
Korean food 181
Korean goods 18, 181
Korea Town: in Osaka 19, 177, 181–2, 184–9, 239, 241
Koreans 17, 18, 31, 116, 177–89, 229–30, 238–9
Kurdistan 170
Kurds 18, 160, 165, 170
Kuwait 167
Kwong, Peter 137

land: allocation of 63, 79; control of 15, 61, 74; cultural and political meaning of 65, 77; market 15; monetary value of 54, 63, 75–6, 105, 139, 178; policy 63, 65, 75, 78; private 66, 74; registration of 165; rights 61; sale of 105; scarcity of urban 10; state control of 66, 70
landownership 7, 9, 11, 63, 77–8, 89; 164; concentration of 63
landscape 144, 156, 189, 209, 232, 240
language 130, 132, 137–8, 150, 169–70, 186; inability to speak local 130, 137; *see also* bilingual people
Laotians 35
latifundium 110
Latin America 122, 213, 219
Latin Americans 27, 31, 228
Lebanese 110
Lebanon 15, 82–5, 87, 89, 92–3, 237; French Mandate in 84; *see also* Beirut
Lefebvre, Henri 6–8, 92, 240
Leisch, Harald 16, 231, 238
Levin, J.S. 47
lifestyle 129, 143–4, 156
Lippo Karawaci 105–6
Little Triangle (Israel) 61–3, 65–6, 68–78, 227, 232, 237
living together with difference (civic ideology in Japan) 19, 177, 183–4, 186–9, 239
local administration *see* government
loneliness 168
Los Angeles 153–4, 230, 232
Low, Setha 81
lower class 2, 16, 39–40, 106, 111–13, 118, 139, 165, 217, 231; lower-class neighbourhood 16, 39–40 (*see also favela*; slums)
Luz, Hélio 124 n. 5, 125 n. 22
Lydda 65

macro-level of analysis 5, 10–11, 13, 18, 20, 25, 236
mainland China *see* People's Republic of China
Malaysia 156
male 123
Manaus 110
mapping 36, 53, 120, 179, 185
markers of social distinction 121
markets 76, 90
Markham 56
Maronites 83, 84, 86–7
Marschalek, Ilse 198, 201
Marxist paradigm 7, 230
Mashhad 166–7
mass murder 111–2
McGee, Terry G. 5, 16
Mecca 167
Medan 102–3
Mediterranean 70, 84
Mediterranean style of architecture 241
meeting place for different ethnic groups 18–9, 92, 196, 200–2, 206–7, 235
mega-urban regions 2, 232
Melbourne 51
melting pot *see* ethnic melting pot
mental map 99–100, 103, 115, 117, 120–1, 197, 205, 231; *see also* perception of urban space
merchants association 181–2, 184–6
meso-level of analysis 10, 13, 17–18, 20, 97, 236
mestizos 210
Mexican–American War 210
Mexicans 154, 210–1
Mexico 210, 219
Michigan 50
micro-level of analysis 5, 10–11, 13, 18, 20, 175, 175, 228, 236
middle class 2, 40, 106–7, 111–12, 114–17, 118–19, 121–3, 149, 161, 164–5; middle-class neighbourhood 40–1
Middle East 52, 235
migration 106, 144, 148, 178, 227; migration to the city 84–5; patterns of Chinese overseas 100
military 8, 237; Israeli 66
militia 15, 81–2, 84, 86–8, 235
Minangkabau 242
Missouri 210
mobile phone 115

mobilization of ethnic support 17, 236; *see also* ethnic solidarity
modernization 121, 162, 164, 178
Moharram 166, 171, 172 n. 3, 240
Moluccas 11–13
Monterey Park 153–4, 158 n. 12
mortality: gender difference of 48, 50–1; rates 47–8; risk 51
mosque 12–13, 18, 53–4, 56–7, 166–9, 171
Mostar 227
mourning 169, 171
moving out of ethnic neighbourhood 16–17, 29, 104, 107, 134, 137, 152–4, 211, 228–30
mugger 121
mulattos 17, 111–13, 116–17, 119, 121, 123
multi-ethnic neighbourhood 160, 202
multiculturalism 144, 205; Canadian policy of 28, 42, 46, 57, 238; *see also* pluralistic identity
mural 130–1, 137
Muslims 12–13, 15, 52–7, 65, 68, 84–5, 87, 105, 202, 227, 240; *see also* Moharram; Shiite; Sunni

Nablus 68
Nagasaki 187
Nahal Iron (Wadi A'ra) 68, 73–5
Nasser 91
nation–state 3, 189
naturalization 181, 190 n. 4
Negev 65–6
neighbourhood 39, 88, 91, 97, 105–6, 112, 114, 117–21, 129, 143, 150, 154, 170, 178, 205, 207, 221, 239, 243; socioeconomic quality 27, 29–31, 35, 39–41, 228 (composite index of 14, 33–5, 228)
New Jersey 134, 138
New Mexico 209–11, 214–15, 217–18, 220
New Orleans, French Quarter 216
new towns 16, 66–7, 105–7, 211, 231
New York 128, 145
New Zealand 149
newspapers 109, 111, 130, 137, 153, 185, 190 n. 7, 221
Nice 90
no-go zone 4, 15, 17, 105, 123, 226, 233
non-governmental organization 193, 199, 204, 238

non-places 117
non-white groups *see* visible minorities
North America 2, 13, 52, 57, 226, 242
North Korea 186
North Sulawesi 101
Northern Ireland 242
nouveaux riches 164

Oikodrom 19, 193–8, 200
Orange marches 242
Osaka 18, 177–80, 182, 184, 239, 241
Ottomans 90

Pacific 142–3, 156
Pacific War 178, 180, 184
Padang 242
Pakistan 3, 166–7
Palestine: state 84; *see also* Israel
Palestinians 15, 87, 233
Palo Alto 145, 150, 156 n. 1
pan-Arabism 91
Paraguay 115
Park, Robert Ezra 1, 9, 88
patronage 12
Peace Village 57
pedestrian zone 194
Pennsylvania 138
People's Republic of China 18, 52, 103, 106, 126, 132, 149, 155–6
perception of urban space 99–100, 220, 222, 228; *see also* mental map
Philadelphia 17, 126–39, 227, 229, 231, 236, 240–1; Philadelphia Chinatown Development Corporation (PCDC) 133–4
Philipinos 52
Philippines 3
Phung, N. 130
Pilgrims 219
place of worship 52–3, 56–7, 229, 233, 242; access to 52; spatial distribution of 14, 46, 52; *see also* church; Hindu temple; mosque; Shinto shrine; Sikh temple
place-making 177, 179
planning *see* development, planning; regional planning; urban planning
plural society 234
pluralistic identity 85, 90
Poles 110
pollution 155
Pontianak 102, 104
poor 116, 118, 121–3, 231

population growth 129
Porto Alegre 110
Portugal 156
Portuguese 110
post-colonialism 4
post-industrial economy 7
post-modern urban conditions 8, 203
power structure 221
Price, Tanya Y. 238
private space 115
procession 222
promenading 90–2, 116, 203, 235
property owners 116, 215
public space 12, 76, 89–93, 107, 109, 114–15, 120–1, 143, 171, 186, 192–3, 195–200, 202–4, 206–7, 209, 222–3, 226, 233–6, 240, 242–3; as arena of conflict 214–18, 234; definition of 233–4; exclusion from 217; as meeting ground 18–19, 92, 196, 200–2, 206–7, 235
public–private interests 199
Puerto Ricans 28
Punjabi 3

Qazir Case 76–7, 237
Qing Dynasty 145
Qom 162
quality of inter-ethnic relations 227, 243 n. 1

race 4, 27–9, 31, 42, 84, 111, 121, 123, 127, 136, 154, 209, 212–13, 217–18, 227, 229, 239, 242, 243 n. 2; restrictive immigration policy in Canada 27
racial categorization 219, 239
racial discrimination 30, 42
racial inequality 29
racism 219
railways 211
Ramlah 65
Ranch 99 megastores 18, 142–3, 151–3, 240
rape 122, 231
real estate 132, 228, 241
re-emigration 101
Recife 110
refugees 3, 11, 15, 35–6, 84, 86, 129, 132, 219
regional planning 62, 75, 77–9
religion 30, 46–52, 84, 219; denomination 84, 93; and social support 47; *see also* spirituality; spiritual needs
religious attendance 48, 51–2, 58 n. 5
religious communities 85, 220
religious homogenization of space 85–6
religious identity 81, 83–4
religious minority 46, 52, 54, 57
remittances 166
rents 92
representation of urban space 209, 214–15, 222–3
residential area 5, 52, 114, 118, 121, 135, 145, 214, 220
residential desegregation 238–9
residential experiences 146–7, 149, 157 n. 6
residential mobility 127
residential segregation 4–5, 9–10, 14–15, 27, 29, 30–1, 34, 38, 42, 54, 79, 81–2, 85, 102, 104, 106, 123, 127–8, 138, 211, 226–30, 232–3, 236, 238–9, 242, 243 n. 1; forced by the government 106–7, 116, 232; index of 32; *see also* condominium; gated communities; multi-ethnic neighbourhood; religious homogenization of space
resistance 17, 139, 187, 229–30
restaurants 128, 130–2, 134, 181, 199, 215–16, 220–2
Rial, Carmen Sílvia de Moraes 16
Rio de Janeiro 109–12, 115, 117–22, 123 n. 3, 124 n. 4, 125 n. 18, 232–3
Río Grande 210
roadblocks 15
Ross, Allen 64
rural-to-urban migration 11, 160, 162, 164, 167, 178, 239,
Russia 122
Russians 61, 64, 67, 75

sacred place 222
Salvador 16, 110, 118
San Francisco 18, 152
San Jose 142–3
Santa Fe 210, 214, 216
Santa Tereza 115, 121
São Paulo (city) 16, 109–10, 112, 114–20, 124 n. 8, 124 n. 15, 232
São Paulo (state) 111, 124 n. 6
Saudi Arabia 3
science park 145
Scott, Allen 63

Sea of Japan 178
security 104–5, 107, 114, 121–2, 124 n. 9, 124 n. 10, 126, 231; guards 118
self-determination of ethnic groups 132, 221; right to 17
self-representation 223
semiotics of space 8
Sennett, Richard 202, 232
sex workers 121
Shah, Chandrakant 14, 18, 229, 238
Shah, Fath Ali 162
Shah Pahlavi II 165
Shahshahani, Soheila 18, 227, 232
Sharon, Ariel 61
Shiite 83, 84, 165
Shinto shrine 178–80, 185, 190 n. 10
Shiraz 160, 166
shopkeepers 214–16, 218, 222; association 180
shopping 149, 151, 153, 166–71; centre 117, 142; mall 16, 18, 104, 106, 132, 153, 194; street 18, 177, 180–3, 185–7
Shoqal Abad 164–5, 167–70, 172
Sikh temple (*gurudwara*) 54
Sikhs 52–5
Silicon Valley 17–18, 142–6, 148–56, 230, 240
Simmel, Georg 117
Singapore 101, 156, 238
Six Day War of 1967 65, 237
slavery 110, 116; in prisons 112
Slovakians 194
slums 119, 121, 123, 124 n. 4, 134, 178–9, 231; slum-dwellers 111, 118; *see also* favela
socioeconomic class 2, 7, 16–18, 39, 41, 82, 92–3, 107, 109, 113, 118, 162, 171–2, 211, 239–40, 242; and ethnicity *see* ethnicity and social class; identity 149–50; indicators 40; mobility 29, 227; *see also* elite; lower class; middle class; neighbourhood; upper class
socioeconomic (in)equality 28–9, 42, 99–100, 110, 122–3, 156, 170, 230; composite index of 33, 35; in Israel 78;
socioeconomic status 16, 29–30, 34, 123, 127, 129, 138, 164, 184, 203
SOLIDERE 15, 88–9, 91–3, 94 n. 8, 235, 237
South Africa 1, 119, 239; Group Areas Act of 1950 239

South Asians 31
South China Sea 102
Southeast Asia 5, 16, 239
Southeast Asians 14, 27, 30–1, 34–42, 129, 132
space 243; of flows 2; meaningless 194; as means of production 6; as object for consumption 6; of places 2; of possibilities 192, 196, 198, 206–7; as social product 6, 8; symbolic value of 6–7, 19, 221; *see also* perception; urban space
Spain 156, 210, 213, 219
Spanish 214, 217, 219–20, 222–3
spiritual needs 56, 131
spirituality 14, 46, 52; *see also* religion
square 91, 120, 175, 194–5, 199, 201, 203, 205, 212, 235
squatters 9, 239
Sri Lanka, people coming from 52
stage, city as 197, 202–4
Stanford Industrial Park 145
state 8–9, 15, 63, 117–18, 123, 183, 226, 232, 236–9, 243
strategic group 8
street 11, 57, 90, 113–14, 120, 122–3, 175, 189, 202–3
street festival 204, multi-ethnic 192–3, 207
street gang 1, 12; fight 112, 124 n. 5; *see also* militia
street names 57, 160, 163–4, 189
street theatre 195–6, 198, 203–4, 206
street vendors 94 n. 11, 120, 153, 180, 211, 215–18, 222
subdivision of land 164, 238, 241; design of 147
suburbanization 144, 181
suburbs 17, 106, 131, 133–4, 136, 138, 142, 145, 152–4, 194–5, 197, 203–4, 228, 230, 232, succession (human ecology) 1, 15, 90, 128
Suharto 12, 99, 101, 103
Sunni 83–4, 160, 165
supernatural forces 11
supramacro-level of analysis 236
sustainability 193–4, 197, 201
Switzerland 122
Syria 167
Syrians 84, 90

Taipei 146, 148, 151
Taiwan 17–18, 142–5, 151–3, 155,

231–2, 233, 240–1; *see also* Hsinchu Science Park, Taipei
Taiwanese 17, 142, 144–5, 148, 150–4, 156, 236, 230, 240
Taiyba 70–1, 73, 77
Tajbakhsh, Kian 8
Tangerang 105
taxes 102, 136; on property 67, 116–17, 211
taxi-dance hall 6
Tehran 18, 160–2, 164, 166–7, 171, 228, 232–4, 240
Tel Aviv 67, 69
Temple Mount 61
tenants 34, 89
territory markers 84
Thai 35
The Truman Show 115–16, 124 n. 13
threat 72, 75, 77, 126; *see also* fear
Tokyo 189 n. 2
Toronto 3, 14, 18, 27, 30–1, 36–42, 46, 52–7, 228–9, 230, 232, 240
tourism 211, 215, 217, 219–20, 223 n. 3, 223 n. 5; ethnic tourism 223 n. 5
touristic image 216
tourists 3, 88, 120–1, 127, 131, 187, 197, 214, 216, 218, 222
town–village relationship 106, 145–6
Tran, H. 130
transcultural identity 142–4
transcultural lifestyle 17–18, 149
transnational community 142, 144, 149, 156; social network 148, 151, 156
Treaty of Guadalupe Hidalgo 210
Tripoli 82
Trudeau, Pierre 28
Tsao, Robert 155
Tulkarm 68, 73
turf war 12, 15, 82, 111, 234–5
Turkey 166–7
Turks 19, 160, 170, 194–6, 203, 205–6; *see also* Ottomans
tyranny of intimacy 202

UN Human Rights Commission 112
unemployment rate 119
United Kingdom 4, 29, 65–6
United States 4, 7, 17, 19, 27–9, 42, 46, 48, 50–1, 87, 110, 122, 126–8, 130, 137–8, 142–3, 145, 148–9, 151, 161, 209–10, 216–17, 228–9, 232–3, 236, 238, 240–1
upper class 2, 39–40, 106–7, 111–12, 114–16, 118–19, 121–2, 231; neighbourhood 39–41
urban administrators 6, 9
urban conflict 222
urban design 209, 222
urban developer 57, 105–6, 127
urban development 138; *see also* urban renewal
urban growth 85, 135
urban planners 6, 9, 57, 93, 127, 156, 156 n. 1, 216
urban planning 8, 56, 61, 63, 74, 86, 89, 91, 134, 136, 138, 182, 184–5, 215, 236, 238; of colonial towns 102; for the expansion of Beirut 84; as political redistribution 63; public participation in 63; strategies in ethnically divided societies 63–4, 236–7
urban quarters 84; *see also* ethnic enclave; ghetto; residential segregation
urban renewal 126, 128–9, 132–4, 136, 138, 205
urban services 197–9
urban sociology 1; new 7
urban space 3–10, 20, 117, 192, 194, 197, 205, 209, 226, 231, 236, 242; allocation of 4; as economic resource 222; meaning of 215; as symbol of power 209, 221; *see also* competition, perception
urban zones 5, 15, 25; zoning 8, 56, 131, 135, 212, 228

Vancouver 30
Varameen 164
Vecoli, Rudolph J. 137
Veloso, Caetano 113
Vienna 19, 192–6, 198–207, 228, 235, 238
Vietnamese 35–6, 132, 139, 228, 240
violence 12, 13, 16, 61, 76, 85, 100–2, 106–7, 109, 111–13, 118–19, 120–3, 231, 235, 242; domestic 111, 113, 123, 125 n. 20; unequal distribution of 118
visible minorities in white-dominated societies 2–9, 27, 31, 42, 52, 196, 202, 228, 239
visiting houses 167–71

Wadi A'ra 68, 70, 73, 77; *see also* Nahal Iron
Wahid, Abdurrahman 101–2

Washington D.C. 137, 238
Wasserweltplatz 193–5, 199–200, 205
Waterworld Festival 19, 192–8, 200–7, 238
weddings 164, 172
West Bank 61, 65–6, 68, 72–3
West, Cornell 28
whites 14, 17, 30–2, 34, 36–42, 48, 81, 111–12, 116, 118–19, 121–3, 130, 154, 212, 228–30, 232, 238–9; definition of 17, 229–30; hegemony 27, 29, 116; ideology of supremacy 28–9, 223, 228–9, 242; neighbourhood 14; *see also* Anglo–Americans
Wirth, Louis 1, 5, 8, 17
women 109, 113, 117, 121–3, 124 n. 3, 125 n. 20, 129, 167–8, 170–1, 232

working class *see* lower class
World Bank 204
World War: I 178; II 128–9, 212; *see also* Pacific War

Yiftachel, Oren 64, 236–7
Yokohama 187
youngsters 91, 112, 122–3, 132, 166, 195, 200–3, 204, 206, 232
youth groups 131
Yugoslavia 227
Yugoslavs 194, 200

Zalafeh 68
Zambia 9, 240
Zionist Congress 66
Zoroastrians 52